SPEEDPRO SERIES

HOW TO CHOOSE
CAMSHAFTS
& TIME THEM FOR MAXIMUM POWER

DES HAMMILL

VELOCE PUBLISHING PLC
PUBLISHERS OF FINE AUTOMOTIVE BOOKS

Visit Veloce on the Web - www.veloce.co.uk

With thanks to my wife, Alison, for her continuing assistance in the preparation and proof-reading of my manuscripts.

Other books from Veloce -

• **SpeedPro titles** •
4-Cylinder Short Block, How To Blueprint & Build For High Performance by Des Hammill
A-Series 998 Engine, How to Power Tune the BMC/BL/Rover by Des Hammill
A-Series 1275 Engine, How to Power Tune the BMC/BL/Rover by Des Hammill
Alfa Romeo Twin Cam Engines, How To Build & Power Tune by Jim Kartalamakis
Camshafts, How to Choose & Time for Maximum Power by Des Hammill
Cylinder Heads, How To Build, Modify & Power Tune by Peter Burgess
Distributor-type Ignition Systems, How To Build & Power Tune by Des Hammill
Fast Road Car, How to Plan & Build a by Daniel Stapleton
Ford SOHC 4-Cylinder 'Pinto' & Cosworth DOHC Engines, How To Power Tune by Des Hammill
Harley-Davidson 1340 Evolution Engines, How To Build & Power Tune by Des Hammill
MG Midget & A-H Sprite, How To Power Tune (New Edition) by Daniel Stapleton
MGB (4cyl), How To Power Tune by Peter Burgess
MGB, How To Give Your MGB V8 Power by Roger Williams
Sportscar & Kitcar Suspension & Brakes, How To Build & Modify by Des Hammill
Weber & Dellorto DCOE & DHLA Carburetors, How To Build & Power Tune by Des Hammill
V8 Short Block, How To Blueprint & Build For High Performance by Des Hammill

• **Colour Family Album titles** •
Citroen 2CV: The Colour Family Album by Andrea & David Sparrow
Citroen DS: The Colour Family Album by Andrea & David Sparrow
Bubblecars & Microcars: The Colour Family Album by Andrea & David Sparrow
Bubblecars & Microcars, More!: The Colour Family Album by Andrea & David Sparrow
Fiat & Abarth 500 & 600: The Colour Family Album by Andrea & David Sparrow
Lambretta: The Colour Family Album by Andrea & David Sparrow
Mini & Mini Cooper: The Colour Family Album by Andrea & David Sparrow
Porsche: The Colour Family Album by Andrea & David Sparrow
Scooters, Motor: The Colour Family Album by Andrea & David Sparrow
Vespa: The Colour Family Album by Andrea & David Sparrow
VW Beetle: The Colour Family Album by Andrea & David Sparrow
VW Bus, Camper, Van & Pick-up: The Colour Family Album by Andrea & David Sparrow
VWs, Custom Beetles, Bugs & Buggies: The Colour Family Album by Andrea & David Sparrow

• **Other titles** •
Alfa Romeo Giulia GT & GTA Coupe by John Tipler
Alfa Romeo Modello 8C 2300 by Angela Cherrett
Bentley Continental, Corniche & Azure by Martin Bennett
British Cars 1895-1975, The Complete Catalogue by David Culshaw & Peter Horrobin
Bugatti 46 & 50 - The Big Bugattis by Barrie Price
Bugatti 57 - The Last French Bugatti by Barrie Price
Caravans, British Trailer Caravans 1919-1959 by Andrew Jenkinson
Chrysler 300 by Robert Ackerson
Cobra - The Real Thing! by Trevor Legate
Datsun Z, from Fairlady to 280Z by Brian Long
Daimler SP250 (Dart) V8 by Brian Long
Fiat & Abarth 124 Spider & Coupe by John Tipler
Fiat & Abarth 500 & 600 by Malcolm Bobbitt
Ford Cortina, Ford's Best-seller by Graham Robson
Ford F100/F150 Pick up by Robert Ackerson
Lea-Francis, The Story by Barrie Price
Lola T70 (New Edition) by John Starkey
Lola - The Illustrated History 1957-77 by John Starkey
Mazda MX5/Miata Enthusiast's Workshop Manual by Rod Grainger & Pete Shoemark
Mazda MX-5/Miata - Renaissance Sportscar by Brian Long
MGA - First of a New Line by John Price-Williams
Mini Cooper - The Real Thing! by John Tipler
Nuvolari: When Nuvolari Raced ... by Valerio Moretti
Porsche 356 by Brian Long
Porsche 914/914-6 by Brian Long
Porsche 911R, RS & RSR (New Edition) by John Starkey
Rolls-Royce Silver Shadow & Bentley T-Series by Malcolm Bobbitt
Rolls-Royce Silver Wraith, Dawn & Cloud/Bentley MkVI, R & S Series by Martyn Nutland
Schumacher, Michael: Ferrari Racing 1998 by Ferdi Kraling
Singer, The Story by Kevin Atkinson
Taxi! - The Story of the 'London' Taxi Cab by Malcolm Bobbitt
Triumph Motorcycles & The Meriden Factory by Hughie Hancox
Triumph TR6 by William Kimberley
VW Bus by Malcolm Bobbitt
VW Karmann Ghia by Malcolm Bobbitt
Works Rally Mechanic - Tales of the BMC/BL Works Rally Department 1955-1979 by Brian Moylan

First published in 1998. Reprinted 1999. Veloce Publishing Plc., 33 Trinity Street, Dorchester DT1 1TT, England. Fax 01305 268864/e-mail veloce@veloce.co.uk/web
www.veloce.co.uk
ISBN 1 901295-19-2/UPC 36847 00119-2

© Des Hammill and Veloce Publishing Plc 1998. All rights reserved. With the exception of quoting brief passages for the purpose of review, no part of this publication may be recorded, reproduced or transmitted by any means, including photocopying, without the written permission of Veloce Publishing Plc.
Throughout this book logos, model names and designations, etc, have been used for the purposes of identification, illustration and decoration. Such names are the property of the trademark holder as this is not an official publication.
Readers with ideas for automotive books, or books on other transport or related hobby subjects, are invited to write to the editorial director of Veloce Publishing at the above address.
British Library Cataloguing in Publication Data -
A catalogue record for this book is available from the British Library.
Typesetting (Soutane), design and page make-up all by Veloce on Apple Mac.
Printed in the United Kingdom.

Contents

Introduction 5

Using This Book & Essential
  Information 6

Chapter 1: Terminology 8
'Duration' 8
'Phasing' 10
'Duration' and 'Phasing' 10
'Valve lift' 11
'Camshaft lobe lift' 11
'Overlap' 12
'Lift rate' 12
'Valve clearance' 12
Mechanical versus hydraulic
  camshaft profiles 13
'Full lift' 13
'Full lift timing' 14
'Piston to valve clearance' 14

Chapter 2: Choosing the Right
  Amount of Duration 16
270 Degrees of duration 17
280 Degrees of duration 17
290 Degrees of duration 17
300 Degrees of duration 18
310 Degrees, or more, of duration .. 18

Camshaft duration and rpm ranges . 19
Pushrod overhead valve engines 19
Single overhead camshaft engines .. 21
Two valve twin overhead camshaft
  engines 21
Four valve twin overhead camshaft
  engines 21

Chapter 3: Checking Camshafts .. 22
Camshaft straightness 24
Camshaft journal condition 25
Lobe accuracy 25

Chapter 4: Camshaft Timing
  Principles 27
Checking TDC 28
Marking the crankshaft damper/
  pulley 29
Camshaft timing check using the
  full lift position 31

Chapter 5: Camshaft Problems ... 34
Engineering tolerances 34
Camshaft specifications 34
Lobe Phasing 35
Re-ground camshafts 35
Lack of power (camshaft at fault?) .. 37

Chapter 6: Timing Procedure
  - Cam-in-Block Engines 38
Checking for full lift using the lifter .. 39
Finding 'true' top dead centre 39
Checking full lift via lifter move-
  ment 40
Checking full lift via rocker arm/
  valve retainer movement 43

Chapter 7: Camshaft Timing
  Procedure - S.O.H.C. Engines .. 45

Chapter 8: Camshaft Timing
  Procedure - T.O.H.C. Engines ... 48
Two valve per cylinder engines 48
Four valve per cylinder engines 48

Chapter 9: Engine Testing 56
Rolling road 56
Track/road testing 57
Drag strip 58
Engine dyno 58

Complementary SpeedPro Books 59

Index 63

Veloce *SpeedPro* books -

  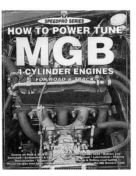 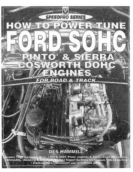 

ISBN 1 874105 82 0   ISBN 1 874105 76 6   ISBN 1 874105 61 8   ISBN 1 874105 81 2   ISBN 1 874105 68 5

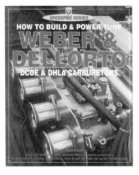 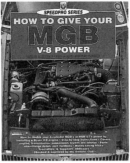 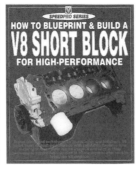 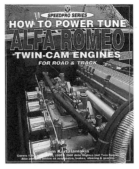 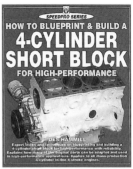

ISBN 1 874105 67 7   ISBN 1 874105 40 5   ISBN 1 874105 70 7   ISBN 1 874105 44 8   ISBN 1 874105 85 5

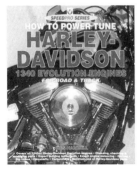   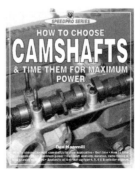 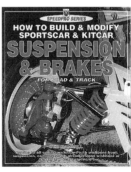

ISBN 1 874105 88 X   ISBN 1 901295 26 5   ISBN 1 901295 07 9   ISBN 1 901295 19 2   ISBN 1 901295 08 7

ISBN 1 874105 60 X   - & there are more on the way

Introduction

This book is designed to give anyone modifying a conventional four-stroke engine sufficiently detailed information to ensure that the right choice of camshaft is made first time - for any individual application. There is also detailed information on how to set up camshafts so that maximum engine efficiency and performance is obtained. All four stroke engines are covered, whether they be older design Siamese port engines, individual port engines, pushrod engines, overhead camshaft engines or twin overhead camshaft engines, two valve per cylinder engines or four valve per cylinder engines, large cylinder capacity, small capacity, motorbike or car engines.

When camshafts are installed in a high performance engine, it should not be assumed that they are correctly timed just because the timing dots line up: there are too many variables for this to be universally true. Much power can be lost if the camshaft timing is out. Also, the camshaft manufacturer's specified settings can sometimes be improved upon if the camshaft is timed a little advanced or retarded from the standard recommended setting.

Note that duration, valve lift and the rates of opening and closing of valves have all been refined to the point that, today, not much development is being carried out in an effort to improve or generate new profiles for the various conventional types of valve mechanisms.

Far too many engines feature camshafts that are not suitable for the particular application and, further to this, camshafts are often fitted with no thought given to the opening and closing points, or full lift timing point, in relation to piston/crankshaft movement.

The camshaft has to be matched to the particular application. There is no point in fitting a 316 degree duration camshaft which develops power between 4200rpm and 7800rpm when, in fact, the engine is going to be turning at between 3000rpm and 6500rpm most of the time. With such a set-up, the engine will not go anything like as well as it should lower down the rpm range (anywhere up to 5000rpm) and will perform no better between 5000rpm and 6500rpm than it would with, say, a good 288 degree duration camshaft.

Des Hammill

Using This Book & Essential Information

USING THIS BOOK

Throughout this book the text assumes that you, or your contractor, will have a workshop manual specific to your engine to follow for complete detail on dismantling, reassembly, adjustment procedure, clearances, torque figures, etc. This book's default is the standard manufacturer's specification for your engine model so, if a procedure is not described, a measurement not given, a torque figure ignored, you can assume that the standard manufacturer's procedure or specification for your engine needs to be used.

You'll find it helpful to read the whole book before you start work or give instructions to your contractor. This is because a modification or change in specification in one area may cause the need for changes in other areas. Get the whole picture so that you can finalize specification and component requirements as far as is possible before any work begins.

For those wishing to have even more information on high-performance short block building principles, ignition systems, Weber/Dellorto sidedraught carburettors and cylinder head work, the following Veloce titles are recommended further reading: *How To Blueprint & Build A 4-Cylinder Short Block For High Performance*, *How To Build & Power Tune Distributor-type Ignition Systems*, *How To Build & Power Tune Weber & Dellorto DCOE & DHLA Carburetors* and *How To Build, Modify & Power Tune Cylinder Heads*.

ESSENTIAL INFORMATION

This book contains information on practical procedures; however, this information is intended only for those with the qualifications, experience, tools and facilities to carry out the work in safety and with appropriately high levels of skill. Whenever working on a car or component, remember that your personal safety must **ALWAYS** be your **FIRST** consideration. **The publisher, author, editors and retailer of this book cannot accept any responsibility for personal injury or mechanical damage which results from using this book, even if caused by errors or omissions in the information given. If this disclaimer is unacceptable to you, please return the pristine book to your retailer who will refund the purchase price.**

In the text of this book **"Warning!"** means that a procedure could cause personal injury and **"Caution!"** that there is danger of mechanical damage if appropriate care is not taken. However, be aware that we cannot foresee every possibility of danger in every circumstance.

Please note that changing component specification by modification is likely to void warranties and also to absolve manufacturers from any responsibility in the event of component failure and the consequences of such failure.

Increasing the engine's power will place additional stress on engine components and on the car's complete driveline: this may reduce service life and increase the frequency of breakdown. An increase in engine power,

SPEEDPRO SERIES

and therefore the vehicle's performance, will mean that your vehicle's braking and suspension systems will need to be kept in perfect condition and uprated as appropriate. It is also usually necessary to inform the vehicle's insurers of any changes to the vehicle's specification.

The importance of cleaning a component thoroughly before working on it cannot be overstressed. Always keep your working area and tools as clean as possible. Whatever specialist cleaning fluid or other chemicals you use, be sure to follow - completely - manufacturer's instructions and if you are using petrol (gasoline) or paraffin (kerosene) to clean parts, take every precaution necessary to protect your body and to avoid all risk of fire.

Chapter 1
Terminology

The most common terms when camshafts are under discussion are 'duration' and 'lift' - almost to the point of exclusion of all other factors. There are several terms that need to be understood so that all relevant factors can be taken into consideration when buying a camshaft or working with camshafts. There are more terms than those mentioned here, but they are not important in the purchase of a camshaft or setting up the camshaft. The terms mentioned here are the ones you really need to know and understand.

There are several terms, and abbreviations of terms, that are used throughout the book from here on. The following abbreviations are to do with the position of the top of the piston in relation to a rotating crankshaft:

TDC - top dead centre (piston at highest point)
BDC - bottom dead centre (piston at lowest point)
BTDC - before top dead centre - (piston rising)
ATDC - after top dead centre - (piston descending)
BBDC - before bottom dead centre - (piston descending)
ABDC - after bottom dead centre - (piston rising)

There are several parts of the camshaft's individual lobes that need to be clearly differentiated between, as each lobe is divided up into distinct areas. They are the 'heel,' the 'nose,' the 'base circle,' opening and closing 'ramps' and the 'flanks' of the camshaft lobe. A camshaft also has 'main bearing journals' and a 'core diameter.' Reference is made to these terms throughout the book, but the 'ramps' and the 'flanks' of the camshaft lobe are the least referred to and the least necessary to know about.

'DURATION'

This is the number of degrees the valves are actually open or 'off their seats' in the four stroke cycle. Although the camshaft rotates at half

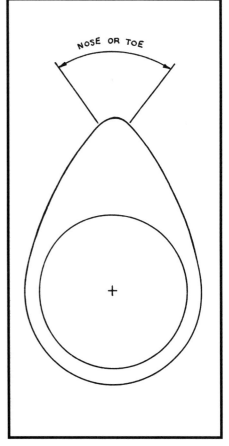

'Toe' or 'nose' of a camshaft lobe.

TERMINOLOGY

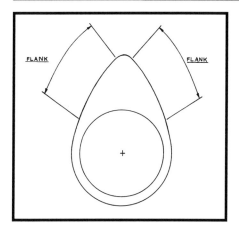
'Flanks' of a camshaft lobe.

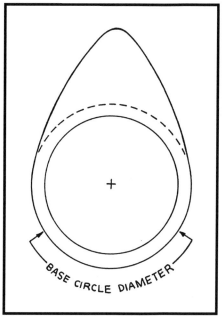
'Base circle' of a camshaft lobe (includes dotted line).

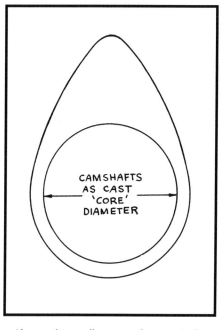
'As cast' core diameter of a camshaft.

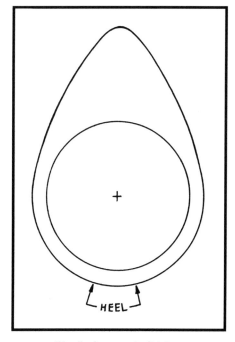
'Heel' of a camshaft lobe.

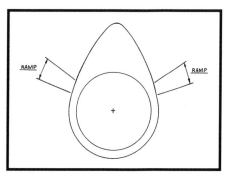
'Ramps' of a camshaft lobe.

Camshaft main bearing journal.

Camshaft's 'as cast' core diameter.

engine speed, the **degrees of camshaft duration are always measured in degrees of crankshaft rotation**. Camshaft degrees are measured at different lifter/follower lift heights so there can be slightly different duration figures quoted for the same camshaft. This is a bit confusing (not to mention annoying) as there is no real set standard in the

SPEEDPRO SERIES

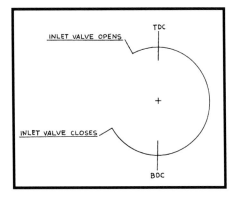

Diagram universally used to show the opening and closing points of inlet valves.

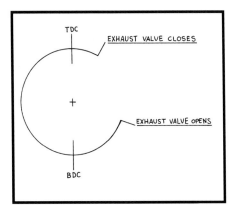

Diagram universally used to show the opening and closing points of exhaust valves.

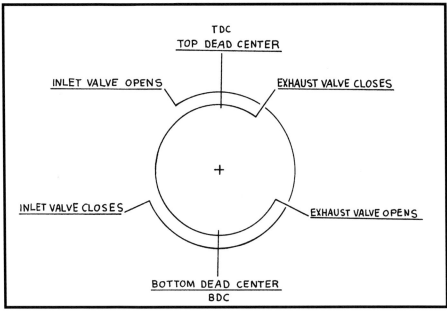

Diagram combining inlet and exhaust timing events. The inner circle, showing the exhaust cycle, is made smaller than the inlet circle.

camshaft industry by which all camshafts can be rated against other similar 'grinds.' It just depends on who measures the duration and where they measure it from. Essentially, though, most people when they are talking duration, mean the number of degrees from the instant that the valve leaves the seat to the instant that it shuts. In the overall scheme of things, as long as you know near enough the just off the seat duration of a particular camshaft, then buying the wrong camshaft for your application will not happen.

The accompanying style of diagram is almost universally used to show the opening and closing points of the inlet valve and the exhaust valve. There is some variation in where the camshaft is measured from on the basis of a 'checking height' but, essentially, this sort of diagram tells you what you need to know and how to rate a high performance camshaft. For convenience, the inlet and exhaust valve opening and closing points are always shown in the same camshaft manufacturer's diagram.

'PHASING'

This is the relationship of the duration of the inlet and exhaust cycles to each other. There are two ways in which the phasing of a camshaft is described.

The first is via the inlet opening and closing points and the exhaust opening and closing points. Even though this method is not always dead accurate, it is the most widely used method of describing what the camshaft's phasing is and is, no doubt, like this because of the universal use of the standard type duration diagram over the past fifty years or so. Basically, we are all used to this method of showing duration and phasing.

The second method of describing the phasing is as a 'lobe centre angle' or the angle between the full lift points of the inlet valves and the exhaust valves. This method is dead accurate, but is less commonly used because it is not so easy to understand and, as a consequence, is not covered in this book to avoid complication. Lobe centre angle is, however, the recommended way of accurately listing the camshaft phasing. The lobe centre angle is effectively the angle formed between the full lift position of the inlet lobe and the full lift position of the exhaust lobe.

'DURATION' AND 'PHASING'

These two factors are always tied together when camshafts are

TERMINOLOGY

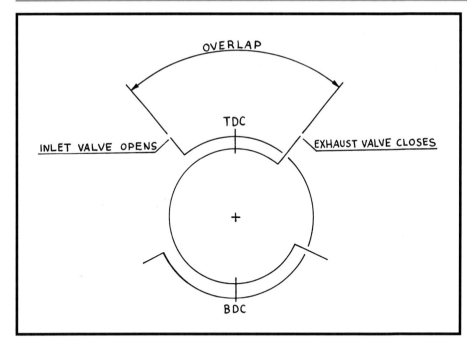

Inlet valve and exhaust valve are both open when the inlet valve is opening and the exhaust valve is closing.

discussed. Duration is the number of degrees that the valves are off their seats (measured in crankshaft degrees) and the phasing is basically the relationship of the opening and closing of the inlet valves and the opening and closing of the exhaust valves (or the relationship of the full lift of the inlet cycle to the full lift of the exhaust cycle). There is some confusion here with duration when camshaft's lobes are measured at slightly different opening and closing heights, but there is only ever a relatively small amount of discrepancy between camshaft manufacturers and grinders. These two factors are, along with valve lift, the universal measure of the camshaft and quite rightly so.

'VALVE LIFT'

This is the term applied to the maximum amount of lift (or maximum distance the valve travels off its seat). This figure will vary from camshaft profile to camshaft profile, and engine type to engine type.

Most modified production four cylinder and six cylinder engines have a general range of valve lifts that go from 0.375in/9.50mm to 0.550in/14.0mm, while larger engines, such as American V8s, will have a general range of valve lifts of 0.450in/11.5mm to 0.650in/16.5mm. The valve lift is loosely tied into the head diameter size of the valves and, more specifically, the inlet valve head size.

Engines with 1.400in/35.5mm to 1.500in/38.1mm diameter inlet valve will have lift range of approximately 0.395in/10.0mm to 0.475in/12.0mm.

Engines with 1.750in/44.5mm to 1.875in/47.6mm diameter inlet valves will have a lift range of 0.425in/10.7mm to 0.550in/14.0mm.

Engines with 2.000in to 2.250in/50.7mm to 57.3mm diameter inlet valves will have a lift range of 0.475in/12.0mm to 0.650in/16.5mm.

'CAMSHAFT LOBE LIFT'

Just because the valves have a certain amount of lift, it does not mean that the camshaft lobe itself has that same amount of lift. Pushrod overhead valve engines almost always work with a rocker ratio system. For example, this means that while a camshaft lobe may have 0.280in/7.1mm of lift, because it moves the valve via a rocker which has a 1.65:1 rocker ratio, the valve will have approximately 0.395in/10.0mm of actual lift. The rocker ratio varies from engine design to engine design, but the usual range is from 1.25:1 to 1.75:1. Rocker lift ratio can be altered, in certain circumstances, by using aftermarket 'high ratio rockers' which may see the standard ratio increased from 1.25:1 to, say, 1.5:1 - this change alone can offer a significant increase in performance in some applications. Not only does the valve lift higher but also faster per degree of camshaft rotation. Within the limits of mechanical reliability this is a win win situation. There are, however, always mechanical limits to the amount by which the rocker ratio can be increased.

Some single overhead camshaft engines use 'finger followers,' which work with a variable rocker ratio via the fact that the nose of the camshaft lobe wipes across a contact pad (point of actual lobe contact is not the same all of the time). Engines like this basically have non alterable theoretical rocker ratios which usually range between 1.5:1 to 1.7:1.

Some single overhead camshaft engines, and all twin overhead camshaft engines (or twin overhead

SPEEDPRO SERIES

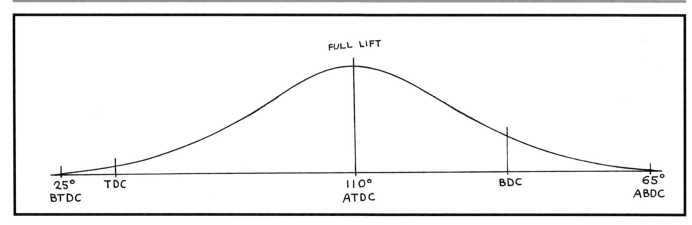

Typical rate of lift for an inlet valve. This example represents a 270 degree duration camshaft with 25 - 65 - 65 - 25 phasing.

camshafts per bank of cylinder engines), have direct acting camshaft lobes which means that, whatever the camshaft lobe's rated lift is, that is what the valve lift will also be. Nowadays, these engines can have mechanical (solid) camshaft followers or hydraulic camshaft followers, but they always used to be mechanical.

'OVERLAP'

Overlap is the number of degrees of duration during which the inlet valve and the exhaust valve are open together (exhaust valve closing, inlet opening). In the four stroke cycle of any engine, as the exhaust valve is closing the inlet valve is opening. The amount of overlap is what causes roughness at idle or low speed.

'LIFT RATE'

This relates to the speed with which the valve is lifted off its seat per degree of crankshaft rotation and, conversely, the slowness with which the valve is returned to its seat. The lift rate varies markedly from camshaft to camshaft. Most performance grinds lift the valve as quickly as possible within the limitations of the valve gear design, and close it as quickly as possible after leaving the valve open for as long as possible. Generally, as the valve is lifted off its seat, the most rapid rate of lift per degree of crankshaft rotation occurs between 30 to 80 per cent of total valve lift.

In the first 30 per cent of a valve's lift, the valve is accelerated from being stationary up to the highest rate for the particular camshaft profile and then slowed down as the valve approaches full lift in the last 20 per cent of the lift phase. At full lift, and about 5 degrees of crankshaft rotation either side of the full lift point, there is very little valve movement taking place.

Valve lift is slowed down as the valve approaches full lift to ensure that the valve follows the camshaft lobe's profile exactly. The accelerative effect of the lift rate, coupled with inertia of the valve and valve train componentry would, depending on the rpm of the engine, see the valve continuing on lifting past the lobe's basic profile if this deaccelerative factor was not taken into account in the design process.

As the valve is closed, the opposite effect of the opening phase is required. The valve is at first lowered slowly from the full lift point, then closed more quickly, then slowed down again as it approaches closing point. In each phase (opening and closing), the idea is to have the valve off its seat for as long as possible, get the valve to full lift as quickly as possible and keep the valve open in the vicinity of full lift for as long as possible (all mechanical limitations being taken into account).

'VALVE CLEARANCE'

This only applies to mechanical (those used with solid, rather than hydraulic, lifters) camshafts and relates to the clearance requirement between the camshaft lobe and the lifter, rocker or camshaft follower. This clearance varies from camshaft to camshaft, and sometimes exhaust lobe to inlet lobe on the same camshaft, and is designed to effect smooth and quiet operation.

On mechanical camshafts a 'quietening ramp' or 'opening ramp' and 'closing ramp' take up the clearance from the 'heel' of the camshaft lobe until there is no gap, and contact between the lifter or follower occurs (resulting in valve lift). Conversely, when the valve is closing the valve contacts the seat and stops, and the gap between the camshaft lobe and the lifter or follower opens gradually.

TERMINOLOGY

MECHANICAL VERSUS HYDRAULIC CAMSHAFT PROFILES

There are subtle differences between hydraulic camshaft profiles and mechanical (solid lifter) camshaft profiles, but the overall effect of both camshafts is very similar. Modern hydraulic camshaft profiles are very good and produce, for all intents and purposes, identical power to their mechanical counterparts. The mechanical camshaft profile has a slight accelerative design feature advantage which, if used to the full, can see the mechanical camshaft equivalent of a hydraulic camshaft profile (we are talking about duration here) have a more rapid lift rate per degree of crankshaft rotation. This all adds up to the valve opening and closing at the same time on each type of camshaft but, if the valve is opened quicker by the mechanical camshaft profile, it stays in the full lift position for longer. This feature is what has lead to the saying that mechanical camshafts have more 'snap' than hydraulic camshafts.

The 'snap' factor should equate to better volumetric efficiency and would, if all things were equal, but there are other factors involved. Things like cylinder head design and limitations which can nullify or limit the efficiency of any camshaft and which, in the end, can mean that the engine goes much the same with either type of camshaft installed.

The slightly better volumetric efficiency afforded by the lobe design of some mechanical camshafts can result in the engine being a bit more willing when the throttle is opened, and producing higher rpm (optimum volumetric efficiency is maintained further through the rev range). If all other considerations in the particular engine are conducive to allowing better volumetric efficiency, then a mechanical camshaft will definitely allow the engine to go better. Ultimately, the only way to find out whether a mechanical camshaft offers a performance advantage is to try the same engine with both types of camshaft.

A guide to whether or not the mechanical camshaft advantage is being exploited is the amount of valve clearance recommended. Some of the more 'wild' American camshafts used in pushrod V8s have valve clearances of 0.024in/0.60mm-0.030in/0.85mm, instead of tighter valve clearances of 0.012in/030mm-0.016in/0.40mm. On smaller four cylinder engines, the valve clearance of an aggressive camshaft will be in the vicinity of 0.018in/0.45mm-0.022in/0.55bmm instead of the more usual 0.008in/0.2mm to 0.012in/0.3mm. Of course, the rocker ratio used on the particular engine will have an effect, but these examples do serve to illustrate the overall difference. Some of the more aggressive camshafts are quite noisy, even when adjusted correctly but, in an engine used exclusively for racing, the noise just doesn't matter.

Essentially, mechanical camshafts are better in the sense of faster valve opening and closing (if this feature has been exploited) but they need frequent valve clearance adjustment and tend to be noisy in operation. The hydraulic camshaft, on the other hand, is quiet in operation when the lifters are in good working order and the lobes are all on size (not worn). Excellent performance is available with a hydraulic camshaft, and the fact that a camshaft is hydraulic should not really count against it for high performance use.

Note that hydraulic camshafts do not have valve clearances or a 'long ramp' (they have a very short ramp) as such, although there have been instances in particular racing classes where hydraulic lifters have been used in conjunction with mechanical camshafts all under the guise of a hydraulic camshaft system. In such cases the hydraulic lifter's internal mechanism is at full travel and the unit effectively functions as a solid lifter. This idea was used to get around class rules of having to use hydraulic camshafts in racing engines (and worked very well ...).

'FULL LIFT'

The full lift point of a camshaft lobe is that point in the middle of the camshaft lobe full lift phase. This can usually be calculated by working back through the duration figures using the universal camshaft duration and phasing diagram. The full lift position of the inlet cycle of the camshaft is almost always listed by the camshaft manufacturer/grinder these days. If the full lift inlet timing point is not listed, ask for it before buying the camshaft, or verify that the camshaft lobes are symmetrically ground as, if they are, it's easy to work out the full lift position.

The example timing diagram (page 14) shows a 270 degree duration camshaft (inlet and exhaust lobes) and the camshaft is phased 25 - 65 - 65 - 25. With a symmetrical camshaft profile, the full lift position of the inlet lobe can be calculated by taking the 270 degree duration and dividing it by two which results in 135 degrees. The inlet opening degrees (25) are subtracted from the 135 degrees which results in 110 degrees. 110 degrees is the full lift position of the inlet lobe after top dead centre (ATDC), provided the camshaft lobe is symmetrical.

SPEEDPRO SERIES

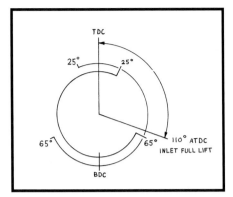

Full lift inlet timing involves finding the full lift point and relating it to crankshaft TDC.

'FULL LIFT TIMING'

Because of the difficulty in picking the precise instant that a valve opens and closes, the full lift of the camshaft lobe point is often used. Within reason it is quite easy to pick the actual full lift point of any camshaft profile. Even if the camshaft profile is a 'long dwell' at full lift type, the precise full lift point can be found using a dial gauge and a degree wheel, or the engine's own crankshaft pulley if it has been marked for full lift timing. This is regarded as being the easiest method of timing a camshaft, but there are reasons why, on some engines, this method of camshaft timing is not used, so it's not a hard and fast rule: note that this is **not** the only camshaft timing method that can be used successfully.

The diagram sequence shows a lifter that is normally associated with a pushrod, but the principle is the same for any camshaft lobe. It is lifter movement that is measured in all cases. The full lift position is determined by relating the lifter distance short of full lift (with an ascending lifter) to the crankshaft degrees at that point, and then continuing to rotate the crankshaft through full lift and on, until the lifter

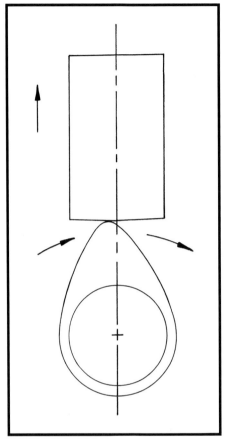

Lifter is ascending as the crankshaft is turned. The crankshaft is stopped when the lifter is 0.010in/0.25mm below full lift. Crankshaft degrees at this point are read off the crankshaft damper/pulley.

has descended to the exact same point again but on the other side of the camshaft lobe.

The full lift point of any camshaft lobe is the central position, in crankshaft degrees, between the ascending lifter stopping point and the descending lifter stopping point.

Whatever timing method is used, all that really matters is that the camshaft is timed so that the individual engine produces the best possible power. The place to start is were the camshaft manufacturer or camshaft grinder has designated; the camshaft can then be advanced or retarded

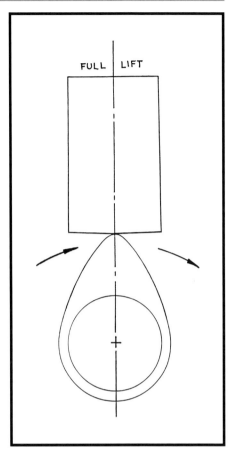

Here the lifter is at full lift: not an easy point to identify precisely, so points each side of full lift are used for accuracy.

(maximum of 4 degrees either way) from this position to see whether power is improved.

'PISTON TO VALVE CLEARANCE'

There are two parts to piston to valve clearance.

The first involves clearance between valves at full lift and piston crowns built into the engine's design. For many engines the amount of built-in clearance will mean that the crankshaft can be rotated 360 degrees, with the camshaft drive disconnected and valve/s at full lift, without a piston crown contacting a valve. The

TERMINOLOGY

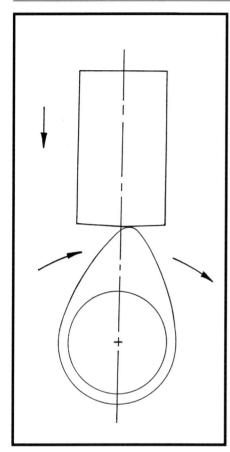

Lifter has now descended to a point 0.010in/0.25mm below full lift as the crankshaft is turned and stopped at this point. Crankshaft degrees for this point are read off the crankshaft damper/pulley.

problem of contact occurs much more on modern engine designs compared to older engine designs (which had deep combustion chambers and valves that were not angled to the bore axis).

Many older, vertical valve engine designs had the valve heads level with the gasket surface of the cylinder head at full lift, meaning that, with the engine assembled, there was the gasket thickness (0.060in/1.5mm) as clearance. In addition, the fact that piston crowns were, invariably, some way (0.020in/0.50mm) down the bore at TDC meant there was almost always 0.080in/2.0mm piston to valve clearance at full valve lift and no chance of the piston and valves ever colliding under any circumstances. Many modern, four valve per cylinder engines, on the other hand, will bend every single valve if each valve is positioned at full lift and the engine rotated. You need to be very sure what sort of engine you have because, if the valves are able to contact the piston, care must be taken when assembling the engine and later when timing the camshaft/s.

In standard form, many engines will be able to lose the drive to the camshaft (chain, belt, and so on) and remain undamaged by piston to valve contact. However, after installing a high performance camshaft this may no longer be the case; the increased lift of the camshaft may now mean that, at full lift, either inlet or exhaust valves could be hit by the pistons if the camshaft drive breaks. If the camshaft drive breaks this still does not necessarily mean that valves will be bent, there's a certain amount of luck involved (none of the valves may be at full lift when the drive breaks). More often than not, however, a valve or two is actually bent and this is precisely the reason that cam belt replacement is recommended at regular intervals. It is certainly much cheaper to replace a belt than a set of valves.

The second part of the piston to valve clearance situation is the 'working clearance' that must exist between the heads of the valves and the piston during every two turns of the engine. This, of course, only applies to engines where the valves can contact the top of the piston at some point when the piston is approaching TDC and there is a certain amount of valve lift present.

In standard form, the engine designer allows for sufficient 'working clearance' so that, in the normal course of events, there is plenty of latitude for error when fitting a new belt drive (in fact, if the engine is set a tooth or two out the valves will often still not hit the piston). This situation changes when the camshaft is changed for a rapid action, long duration, high lift alternative. The valves open earlier, close later and lift higher, and, if the cylinder head has been planed to increase the compression and a thinner head gasket fitted, the working clearance could be reduced by 50% or more. So, where there was originally 0.200in/5.0mm clearance, there could now be 0.090in/2.3mm at certain points in the valve opening and closing cycle. In this situation, timing for safe clearance becomes quite critical, to the point that valve reliefs may have to be cut into the tops of pistons or existing reliefs deepened to improve the piston to valve clearance. It's quite possible to bend valves when assembling some engines, and even easier to bend valves setting the camshaft timing. Having the maximum possible piston to valve clearance on any high performance engine is a well founded principle!

Visit Veloce on the Web - www.veloce.co.uk

Chapter 2
Choosing the Right Amount of Duration

The majority of high performance orientated camshafts have durations of between 270 degrees and 310 degrees. The 270 degree duration camshafts being 'mild' and the 310 degree duration camshafts being 'wild.' This is the generally accepted range of durations for any camshaft on the market today.

Tests have shown, however, that for virtually all performance applications the range of duration is confined to between 270 degrees and 310 degrees (give or take two or three degrees). This includes engines which are going to be used exclusively on the road through to full race engines that are going to be race only. This 40 degree difference between the low side and the high side of the range may not sound a lot but, in practice, it makes a lot of difference.

Note that there have been camshafts that have had 340 degrees of duration, but they represent the absolute maximum and can be considered beyond any form of sensible use. Such camshafts had the inlet valves opening in the vicinity of 68 degrees BTDC and closing 92 degrees after BDC, the exhaust opening 92 degrees BBDC and closing 68 degrees ATDC.

Camshaft duration listings go up in stages of 10 degrees from 270 degrees, but many camshafts have odd figures such as 272, 284, 297 and 312. What this means is that all of the timing events listed are approximations only and rounded off for convenience. The effect of a 272 degree duration camshaft is similar to a 270 degree duration camshaft, while a 288 degree duration camshaft is similar to a 290 degree duration camshaft. Also, two camshafts may be advertised as having the same duration but the phasing could be different on each. The inlets and/or exhausts could open anything from 1 to 3 degrees earlier (or later) on one camshaft and close 1 to 3 degrees earlier (or later).

In the interests of clarity, all camshaft durations listed in this book have the inlet opening point (in degrees) first followed by the inlet closing point (in degrees) and then the exhaust opening point (in degrees) followed by the exhaust closing point (in degrees). So when a camshaft timing is listed as being 42-78-78-42, the first 42 is the point in degrees that the inlet opens BTDC, the first 78 is that point in degrees that the inlet valve closes ABDC, the second 78 is that point in degrees that the exhaust valve opens BBDC and the second 42 is that point in degrees that the exhaust valve closes ATDC. All camshaft duration measurements are in **crankshaft** degrees.

The figures listed in this book are what are termed 'absolutes' as they are taken when the camshaft lobe just starts to impart actual valve lift (basically, that first 0.001in/0.025mm-0.003in/0.075mm of valve movement). Although there are many

CHOOSING THE RIGHT AMOUNT OF DURATION

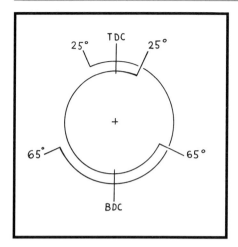

Typical phasing of a 270 degree camshaft.

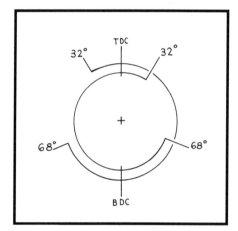

Typical phasing of a 280 degree camshaft.

variations on how to rate camshaft durations using different lifter rise heights, and so on, this is the way that most people measure camshaft duration.

Some camshafts durations are advertised at a certain lifter rise. For example, Harley-Davidson motorcycle camshafts are all pretty much standardised at 0.053in/1.35mm lifter rise, irrespective of who grinds the camshaft. Many aftermarket automobile engine camshaft manufacturers use lifter rises of 0.020in/0.5m which is confusing for most people when their catalogues are read. The manufacturers are actually trying to be helpful by giving duration degrees at a certain lifter rise in the interests of accuracy but, in the end, most people are familiar with 'absolutes' and end up wanting to know what the 'real' number of degrees is!

All camshaft durations listed are shown in similar diagrams and these also show the phasing. This gives a clear picture of how duration and phasing are related, and the parameters of camshaft duration.

270 DEGREES OF DURATION

Consider 270 degree duration camshafts to be the mildest camshafts available, but in no way is it to be inferred that they are inferior because they are of mild duration. These mild camshafts are frequently similar in duration (sometimes less) to the standard camshaft. When such camshafts are fitted to a modified engine, they will almost always perform better than the standard type camshaft that they replace because they give a faster rate of lift off the seat to full lift, hold the valve in the vicinity of full lift and then return the valve to its seat at a fast rate: this means the valves are off their seats for as long as possible within the 270 degrees of duration. There is nothing lazy about the valve action of these cams.

The timing events for a 270 degree camshaft will be phased in the vicinity of 25-65-65-25. These camshafts will have a smooth idle (only just) and are recommended for all-round road use where good fuel economy is desirable and emission levels must be acceptable.

280 DEGREES OF DURATION

280 degree camshafts are the next step up and offer an improvement in performance. Although this amount of duration is still considered mild, camshafts with this amount of duration which have been fitted to a modified four valve per cylinder engine, will produce excellent power to 7500rpm. The efficiency of the cylinder head accentuates the effect of the camshaft to a very marked degree. What makes an engine go is good volumetric efficiency or cylinder filling (plenty of air/fuel mixture in the cylinder).

The timing events of a 280 degree camshaft will be phased in the vicinity of 32-68-68-32. With this amount of duration the engine will not have a dead smooth idle, and such camshafts can be regarded as the start of high performance camshafts in the true sense.

290 DEGREES OF DURATION

The timing events of a 290 degree camshaft will be phased in the vicinity of 37-73-73-37. The cylinder head plays a considerable part in the effect of a camshaft. A standard engine that is fitted with a long duration camshaft like this is seldom efficient (produces good power). The reason for this is that, even though the valves are opened earlier and perhaps higher, the actual inlet tract is completely standard and, as a consequence, is the limiting factor to volumetric efficiency. This is nothing to do with the camshaft, but it frequently limits the effect of the camshaft timing. These camshafts are used when good mid-range power is essential.

These camshafts give lumpy or rough idles. They generally perform much better in many engines than might be expected when the induction side of things is correctly modified (porting, and so on). 290 degrees of duration is not too little for many applications. It is often said that the

SPEEDPRO SERIES

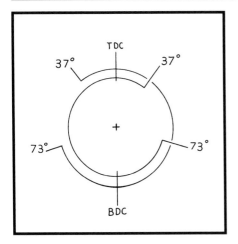

Typical phasing of a 290 degree camshaft.

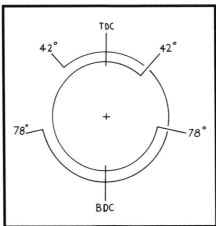

Typical phasing of a 300 degree camshaft.

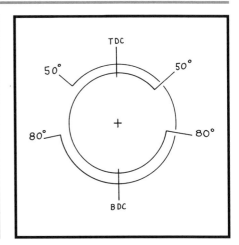

Typical phasing of a 310 degree camshaft.

camshaft is the heart of the engine, but this is only true if all other factors are taken into consideration such as induction and exhaust manifolding. Any engine can be 'cammed up' but if, for instance, the cylinder head is relatively inefficient, there will be no improvement in performance: in fact, power can be lost.

300 DEGREES OF DURATION

The timing events of a 300 degree camshaft will be phased in the vicinity of 42-78-78-42. The degrees of overlap are increasing (that is the two 42 degree figures in this case, and not the 78 degree figures) and it is this increasing overlap that will cause an engine to have a rough idle. 300 degree camshafts are racing camshafts and all engines equipped with camshafts like these will have very rough idles. The 'wilder' the camshaft the higher up the rpm scale the power band starts.

Many two valve per cylinder engines have volumetric efficiency limits which mean that, irrespective of how much duration the camshaft has, you will be getting into a situation of diminishing returns. It's quite possible to fit a camshaft with so much duration that the production of power starts very high up in the rpm range but the efficiency of the cylinder head is so limited that the effective power band is reduced to an unusable level. A better option would be to reduce the camshaft duration so that power is produced lower down the rpm range, so giving the engine a much wider and more usable power band.

310 DEGREES, OR MORE, OF DURATION

The timing events of a 310 degree camshaft will be phased in the vicinity of 50-80-80-50. At this point the 80 degree figures have not increased by very much. The reason for this is that opening the exhaust valve earlier than 80 degrees BBDC is about the limit, and closing the inlet valve any later than 80 degrees ABDC is about the limit for any engine. Opening the exhaust valve at 80 degrees BBDC means that there has been 100 degrees where the combustion pressure has been working in the cylinder. Closing the inlet valve at 80 degrees ABDC allows for 100 degrees of crankshaft rotation before TDC. For many engines this sort of timing represents the maximum that they can respond to and can be considered for racing only. Longer duration camshafts can be fitted, but very often the engine just does not go any better and power production starts later (instead of 3500rpm it may well be 4500rpm).

Camshafts with timing events of a 324 degree, for example, will be phased in the vicinity of 58-86-86-58. This sort of camshaft duration, or more, represents the maximum that is used in the highest revving race engines. That is 9000-10000rpm plus, with the power effectively starting at 5000-6000rpm, or so. No suggestion is being made here that these long duration camshafts do not work, just that they are not practical for most applications.

Note that, with these long duration camshafts, the degrees of duration do not increase proportionately on the inlet closing point or the exhaust opening point. The example camshaft has the inlet valve opening at 58 degrees BTDC and the exhaust valve closing at 58 degrees ATDC. This is 118 degrees of overlap where the inlet and exhaust valves are open together. This is a lot

CHOOSING THE RIGHT AMOUNT OF DURATION

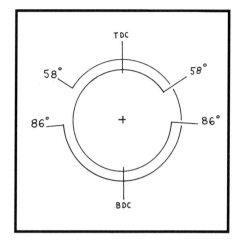

Phasing of a typical 324 duration camshaft.

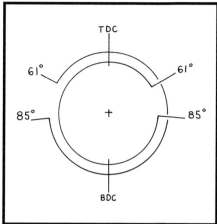

Timing events of the DA1 camshaft.

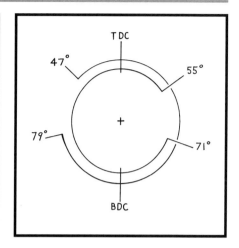

Timing events of the L1 camshaft.

of overlap by anyone's standard, and the idle will be very rough and low speed tractability near non-existent. Another consideration is, of course, the loss of torque throughout the rev range because of the early exhaust valve opening. The pressure in the cylinder is still very high at this point and still pushing hard on the piston (producing torque), and, if the valve is opened too early, torque is lost. The converse of this is that if the exhaust valves are opened too late, the cylinder will not be cleared before the piston starts to rise and engine efficiency will be lost, but for a different reason.

Of course, in some instances, the very low speed performance of a very long duration camshaft-equipped engine just does not matter, but the mid-range power of engines equipped with camshafts like this is seldom as good, in the final analysis, as engines equipped with slightly milder camshafts. Be prepared to change camshafts if the mid-range power is not good enough for the application.

An example of the value of shorter duration camshafts is a converted to natural aspiration four valve per cylinder Ford Sierra Cosworth engine (normally turbocharged) which can be reliably revved to 8500-9000rpm in racing trim. If such an engine is equipped with DA1s (timed at 61-85-85-61), these camshafts will allow the engine to produce top end power at 9000rpm second to none.

The same engine equipped with L1s (timed 47-79-71-55) will pull much harder (more torque) from 5000rpm through to 8500rpm but, ultimately, produces less maximum power than the DA1-equipped engine. The L1-equipped car is, in many instances, much quicker than the DA1-equipped engine car (same car/engine with camshaft changes only) and certainly easier to drive for most people.

CAMSHAFT DURATION AND RPM RANGES

The range of camshaft durations already described will, one way or another, suit virtually all engines. The next stage in narrowing down the duration requirement is the required rpm range of the particular engine. This factor almost always is dependent on the volumetric efficiency capability of the cylinder head and then, more specifically, the inlet ports or inlet tracts as a whole. This means that the rpm capability of various camshaft durations can be narrowed down by the design of the cylinder head. The effectiveness of each amount of duration and phasing of a camshaft is almost always going to be determined by the cylinder head design and cylinder head efficiency. With the maximum rpm/maximum cylinder head efficiency known, the rpm range of the various camshaft durations can be equated.

Pushrod overhead valve engines

Engines in this category have the traditional pushrod-operated valve system as used for years on engines like the BMC A-Series, the Ford CrossFlow and pre-CrossFlow, the small block Chevrolet and Ford V8 engines, for example. Most of these engines have cylinder head designs which have a basic efficiency limitation beyond which, no matter how much the standard cylinder head is modified, the power production will not improve. The maximum rpm most of

SPEEDPRO SERIES

these engines are efficient to is around 7500rpm. All of these engines will rev more than this, of course, and frequently do, but power is not increasing at these higher revs, it is diminishing. The duration of camshafts used to achieve 7500rpm, and above, is usually between 300 and 310 degrees, and the cylinder heads have to be well-modified to get the power surge to go this far up the rpm range. If a realistic rpm range is chosen for a particular engine and application, a realistic camshaft duration can be chosen.

There are instances where engines, such as 5-litre V8 engines, for example, are fitted with long duration camshafts (315 to 325 degrees) which have very high valve lifts and very aggressive lobe actions (fully rollerised valve trains). Such engines are used with very narrow power bands (6000rpm to 7500rpm) and produce excellent power (100 to 105bhp per litre) but they need plenty of gears because, under 5500rpm, torque is well down and they just don't 'go' out of their narrow power band. This is a racing only situation, and a specific one at that.

The effective power bands given are approximations only, because all engines vary in design and have different limitations. Some engines, when modified, will go exactly to the rpm listed but some will not. Some of the larger capacity engines or, more specifically, larger cc per cylinder engines with heavy internal componentry, may not make the listed revs.

Effective power bands
270 duration 1000-6000rpm
280 duration 2500-6500rpm
290 duration 3200-7000rpm
300 duration 3800-7500rpm
310 duration 4300-7700rpm

Small four cylinder inline engines, like the A-Series Mini which have 998cc or 1275cc, for instance, respond best to camshafts in the 300 degree range when maximum torque/power is required. The Austin/Rover designed 648/649 camshaft is, overall, the best camshaft available for these engines. This camshaft has 300 degrees of duration, a rapid lift rate and 50-70-75-45 phasing, and, with 1.5:1 rocker ratio rockers, has 0.470in/11.9mm of valve lift. There are longer duration camshafts available but, overall, they seldom seem to make the engine go much better, if at all (cylinder head efficiency limitation).

At the other end of the scale, American small block V8 engines equipped with conventional lifters, and standard cylinder heads which have been well-modified, all seem to produce maximum power between 7200 to 7500rpm. To do this the camshafts need to have at least 300 degrees of duration, with typical timing being 42-78-78-42 and valve lifts between 0.475in/12.1mm-0.550in/14.0mm. If more duration is used the limit is 310 degrees with typical timing being 50-80-80-50 but, really, this is too much duration and mid-range power will not be very good.

With nearly all of these pushrod-type V8 engines, cylinder head efficiency is what determines the maximum engine efficiency point in the long run. No 300-310 degree duration performance camshaft will produce good power in virtually any of these engines if the cylinder heads are standard. The cylinder heads just cannot flow the necessary air, even though the valve is open long enough, so a very long duration camshaft can be fitted with no benefit and a gain in idle speed roughness and loss of low

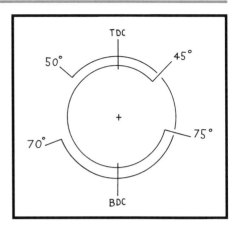

Timing events of a 649 Austin/Rover camshaft.

rpm torque.

The acid test here is, of course, to run one of these engines with standard heads and then have those same heads modified. Provided the heads are modified well, the difference is immediate and you would not think it was basically the same engine. Take, for example, a well-equipped V8 engine capable of being reliably revved to 7200rpm. The engine will now idle very roughly whereas, before, it was just lumpy. The engine will have a different exhaust note (louder) and will really rev freely. The mid-range power will be totally different and the engine will pull through to 7200rpm easily, whereas, before, the engine was reluctant to go over 5500-5700rpm.

In many instances 300 degrees of duration is simply too much. If the engine is basically standard equipped, a pushrod V8 is not usually reliable at 7500rpm, but it will be reliable up to 6500rpm after careful rebuilding. In such cases, a 280-284 degree duration camshaft can be much more effective because the mid-range power is better. Such a camshaft is likely to have phasing of 32-72-70-34 and will (on a modified head engine) produce power up to and at 6500rpm, and pull well

CHOOSING THE RIGHT AMOUNT OF DURATION

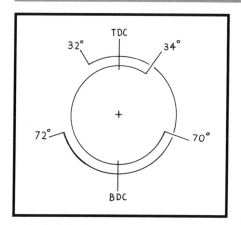

Typical timing events of a 284 degree camshaft.

from off idle (2200rpm or so).

The 310 degree camshaft-equipped modified head engine will not start producing power until 4000-4300rpm, but will not produce much more, if any more, power at 6500rpm than a 280-284-cammed engine. Avoid having a large duration camshaft in an engine revving to 6500rpm maximum, as the effective power band is just too narrow and the mid-range power virtually useless compared to a lesser duration camshaft (280-284 degree) in the same engine.

Single overhead camshaft engines

The majority of these engines can have good volumetric efficiency up to 7500-7800rpm (perhaps 8000rpm on some high/angled inlet port designs) if the cylinder head is well modified. This design of engine will frequently be able to take up to 9000rpm operation reliably as far as the camshaft and rockers are concerned, but the inlet port/inlet tract and exhaust port/exhaust tract will not. It is common for this sort of engine to have a 300 to 310 degree camshaft fitted for all-out racing - and go very well - but rpm and power are limited by the airflow capacity of the head.

Effective power bands
270 duration 1000-6000rpm
280 duration 2500-6600rpm
290 duration 3000-7200rpm
300 duration 3500-7500rpm
310 duration 4000-7800rpm

Two valve twin overhead camshaft engines

These engines are always very efficient, but they cannot match the four valve per cylinder engine. The breathing ability/volumetric efficiency of these engines means that they will produce good power up to 8000-8500rpm if the cylinder head has been well modified, and the internal engine parts are strong enough. The duration of the camshafts need only be 300 to 310 degrees. If the cylinder head is not well modified the rpm ceiling for good power will be less (in some cases as low as 6500-7000rpm).

Effective power bands
270 duration 1000-6500rpm
280 duration 2000-7000rpm
290 duration 3000-7500rpm
300 duration 4000-8000rpm
310 duration 5000-8500rpm

The cylinder head design can be quite a restriction and, when it comes to road-going cars, and you can see why. The designers have often created engines of this design to give a sporty image to the car but know that the last thing the customer really wants is a racing engine. The porting is always designed conservatively so the engine produces good miles per gallon and is lively up to a point which suits most ordinary drivers. Invariably, to get any worthwhile power gains out of most of these engines (unless it is a specifically built racing engine), the cylinder head has to be substantially and well ported.

Four valve twin overhead camshaft engines

The useable rev range alters dramatically with this sort of engine because the cylinder heads are so efficient - especially after they have been well-modified. These engines can have 300 degree duration camshafts fitted to them and produce absolutely excellent power at 9000rpm and still be going strong. Many of these engines in full racing trim have camshafts fitted to them with no more than 310 degrees of duration, but are still producing power at 10,000rpm plus.

Effective power bands
270 duration 1000-6500rpm
280 duration 2500-7500rpm
290 duration 3000-8000rpm
300 duration 4000-9000rpm
310 duration 5000-10000rpm

The duration range for these engines is the same as any pushrod engine and the valve lift likely to be less. The cylinder head efficiency (or volumetric efficiency) of these engines is the factor that allows the cylinders to be filled well at a much higher rpm than a pushrod engine. When the cylinder is not being filled properly, power production slows, or ceases to rise, and this is the point at which power falls off: it's as simple as that.

Virtually all modern four valve per cylinder engines will, when modified, produce excellent power up to 7000-7500rpm with a camshaft change (280 degrees of duration) and 7500-8000rpm with minor porting work with the same camshafts. This sort of duration would never work to this rpm in an older design, vertical valve, pushrod engine.

Chapter 3
Checking Camshafts

Camshafts are precision ground components and, as such, have design specifications which they can be checked against to prove whether or not they are accurately manufactured. There are occasions when things go wrong during the grinding process and, for one reason or another, unserviceable camshafts do end up in engines. The fitting of a less than perfect camshaft can be avoided to a certain extent, but there are parts of a camshaft that can only be checked with complete accuracy in a camshaft grinding machine. The aspects of the camshaft that cannot be easily checked are the lobe profiles and the phasing of the lobes. Nevertheless, any replacement camshaft should be checked before being put into an engine.

When camshafts are ground incorrectly the actual fault can be quite difficult to find. In the final analysis, though, there is nothing on a camshaft that cannot be checked to a fairly high degree of accuracy when either the camshaft is installed in an engine or the camshaft is out of the engine, but checking does take some time.

This lobe is well worn and the camshaft is unusable because it. The toe is well down, in fact by 0.10in/2.5mm.

Except for the actual lobe profile shape (contour), the camshaft can be checked reasonably well before being installed in an engine. Even the lobe profile condition can be roughly ascertained because it takes a lot of use to remove the original grinding marks from the toe (or nose) of a camshaft lobe. In many instances, the loss of correct lobe shape, through wear, can be felt with your fingers. Any camshaft that does have irregularities that can be felt in this manner is not new and certainly not in

Close-up of a very worn lobe.

good condition. In the final analysis, it is quite easy to pick a good lobe from a bad one: wear or damage is almost always easy to see or feel.

There can be some risk when fitting a used camshaft to an engine, but failures of this type can almost always be prevented if the camshaft is thoroughly checked and inspected first. Many used camshafts are, in fact, in excellent condition.

CHECKING CAMSHAFTS

This lobe is hardly worn at all: a camshaft with all of its lobes in this condition is quite serviceable.

Camshaft lobe that has been built up with 'Colmony.'

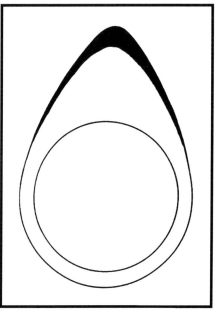

Reground built-up lobe. Built-up area is shaded in black.

Most high performance engines are fitted with new camshafts or reground camshafts and the integrity of the camshaft taken for granted because it is new. There is a considerable amount that can be done - and should be done - to check any camshaft, new or used, to remove all possible doubts about the camshaft's integrity before it is installed in the engine. There's nothing worse than having to remove a camshaft from an engine to track down a fault when, if the camshaft had been checked before being fitted to the engine, it could have been prevented.

Several things can be wrong with camshafts, new or used. For a start, the camshaft can be bent, which will cause it to be hard to turn by hand. The usual problem with used camshafts, however, is lobe wear. The lobes are under considerable stress at full lift because of valve spring pressure. When the lobes do start to wear, the lift they provide reduces and the lift rate per degree of camshaft rotation is also usually down because the lobes wear backward from the toe (nose) and down each side of the lobe (flank). When lobes become worn valve action gets quite lazy and engine performance is lost.

Further to this, not all lobes wear at the same rate and, frequently, one or more lobes wear excessively in comparison to the other lobes on the same camshaft. It only takes one lobe to be well down (0.060in/1.5mm, or more) and the camshaft is virtually a write-off.

Note that camshaft lobes can be successfully repaired. Such lobes can be built up using a hard surfacing agent such as 'Colmony' (British product) which is an easy flowing type of Stellite. If most of the lobes on a particular camshaft are in perfect condition, the faulty ones can be built up and then have their profiles reground. The repaired lobe will almost always never need to be reground again: Colmony is very hard and durable if it has been applied correctly.

In fact, many camshafts with small base circles that are going to have a performance profile put on them have all lobes built up right from the start. This is common on engines with camshafts that have small core diameters, and for which the range of standard adjusting shims is limited (engines such as the old Jaguar XK six cylinder unit come to mind here). The lobes are built up allowing excellent aggressive grinds to be put on the camshafts, but standard shims can still be used to adjust the valve clearances. If the camshaft is done 'on the cheap' and simply ground down (undercut core) it is generally unsatisfactory in the long run as special shims have to be made up, the inverted buckets lift up out of the cylinder head and the whole thing gets messy. Better to avoid this sort of nightmare and pay the high cost of building the camshaft lobes up in the first place; from then on everything is straightforward.

Some camshaft companies rework and sell camshafts which use this built up lobe feature to preserve core diameter. Iskenderian, for example, has a range of camshafts available with what it terms 'hard surface overlay.' They virtually never wear out, but they cost a lot. Such camshafts will take high valve spring

SPEEDPRO SERIES

Camshaft being checked for run-out is mounted between centres in a lathe. Check each journal in turn. The journals could have run-out of 0.001-0.002in but, ideally, there should be no run-out at all.

pressures and show no signs of wear after a considerable amount of use. Lobe damage is possible if a roller lifter fails, though.

CAMSHAFT STRAIGHTNESS

The straightness of any camshaft can be checked between the centres of a lathe, on a camshaft grinder or between Vee-blocks using a dial gauge. Lathes are much more common than camshaft grinders, so it is more likely that this will be the machine tool used for this purpose.

With cam-in-head cylinder heads, which have pairs of removable camshaft bearing shell inserts, the straightness of the camshaft can also be checked using the cylinder head. The front and rear bearings only are left in position and the camshaft is fitted to the cylinder head using two bearings only. The caps are fitted, tightened to specified torque and a dial gauge used to check the run-out of the exposed camshaft main bearing journals.

Caution! - Irrespective of whether the camshaft is new or used, it **must** be straight and checked to be so. The ideal 'run-out' is zero, but a tolerance of up to 0.002in/0.050mm is acceptable on long camshafts. There is

Dial gauge and magnetic stand.

no reason why a camshaft should not be dead straight, irrespective of how long it is, but there has to be some tolerance.

Caution! - Any camshaft which is known to be straight **must** be easy to turn when fitted to the engine block or cylinder head.

For cam-in-block engines, this can easily be checked by simply fitting the camshaft into the block and fitting the thrust plate or thrust washer/s and sprocket so that it is located just as it will be when operating in service, and then turning the camshaft by hand. Any binding means the the camshaft's bearing journals or/and bearings are damaged. Irrespective of what is causing the problem, the time to fix it is **now**. If the bearings are damaged (burred during fitting) they will have to be eased, which can sometimes be done with a three-corner scraper. Bearings that bind the camshaft or, more specifically, that part of the bearing that is binding on the camshaft journal, will be shiny and quite easy to locate. If the camshaft bearing/s cannot be eased *in situ*, they must be removed and either replaced or 'fettled' to remove the error and then refitted. The acid test is that any camshaft must rotate freely in the block when turned with light hand pressure. If a camshaft does not turn in an engine block the problem is invariably one of camshaft journal sizing or damage: the block will seldom be out of alignment (that is, the factory bored tunnel axis in the block).

Camshafts which fit in the head can be checked just as easily by fitting the camshaft into the cylinder head and turning the camshaft to see how easily it rotates. Once again, there must be no binding of any description and, if there is, attention should, initially, be directed at the bearings.

However, there is a further consideration: although cylinder heads are normally very rigid castings, they have to tolerate a considerable amount of heat, especially if the engine has suffered a blown head gasket at some point. What happens then is, of course, that the cylinder head becomes warped and, to make it serviceable, it has to be planed to restore its gasket surface flatness. However, what is often not considered is the fact that, while the head gasket surface has been planed to restore it, the cylinder head is still warped in other areas. The end result of this is that the camshaft axis (as bored in the cylinder head by the manufacturer) can become well out of true (up to and beyond 0.040in/1.0mm) to the point that there is severe binding and difficulty in turning the camshaft.

It is sometimes possible to fit camshafts to warped heads when the camshaft cores are quite small so that the camshaft bends as it is rotated in the engine. The end result of this sort of situation is usually a broken

CHECKING CAMSHAFTS

Correct camshaft journal finish.

Checking a lobe base circle with camshaft mounted between lathe centres. In the area that is circular the permissible run-out is 0.0010in/0.0254mm.

Camshaft lobe being checked via a lifter. The dial indicator is positioned to read vertical lifter movement. It is not advisable to oil the lifter body during this check.

camshaft and bent valves! Such cylinder heads are, basically, write-offs because of the cost of remachining them to restore the camshaft bearing axis. Heads like this have to be align bored with all manner of other remedial work. If the cylinder head is a rare one, or hard to replace, then the necessary work might be carried out, but there will still be associated problems such as restoring the 90 degree axis of the valve and the bores of the followers!

CAMSHAFT JOURNAL CONDITION

Check the surface finish of the camshaft journals. The journals must have a surface finish which is just like that of a crankshaft main or big end journal: this means a mirror finish. Excessive journal roughness will cause cam bearing wear and the rate of wear will depend on the amount of roughness. This is a high load area, especially so when high valve spring pressures are used together with high engine rpm.

Camshaft journals get worn or scored through an engine being run with dirty oil or a component failure somewhere else in the engine from which debris has spread throughout the engine. The dirt particles embed themselves in the surface of the soft material (aluminium of the head or replaceable camshaft bearing) and work away on the journal surface as the camshaft is rotating. The other possibility is that the journals got 'etched' when the camshaft was Parkerized or Phosphate coated. If the camshaft journals are not covered during this process, and the camshaft left in the tank too long, it's possible for the journals to be irreparably damaged. The journal surface can be polished but the roughness will always be there and the bearing will not last in service with strong valve springs. The camshaft journals **must** be dead smooth and polished, **nothing** less will do.

LOBE ACCURACY

Hydraulic (those used with hydraulic followers) camshaft lobes should all be checked for base circle run-out. There must be virtually no run-out (maximum of 0.001in/0.025mm) on a hydraulic camshaft. If a camshaft has one or more lobes with excessive run-out (0.003in/0.076mm, or more) the lifters will not be able to operate correctly as the lifter/follower that is operated by the faulty lobe/s will not allow the valve to shut. Such an engine will have a 'miss' at idle, throughout the rpm range and at full revs.

If each lobe's base circle is checked for run-out there is no doubting the integrity of the component and, if tuning problems arise in the future, the camshaft has already been eliminated from the probable cause list.

It takes five minutes to check a hydraulic camshaft lobe for base circle run-out between the centres of a lathe using a dial gauge and magnetic stand. This check should be made at the same time that the journals are all checked for concentricity. Camshaft lobes can also be checked for run-out in the block, in the cylinder head or mounted on Vee-blocks using a dial gauge and stand, but it is generally easier between the centres of a lathe.

If a camshaft is in the block the run-out (or lack of it) can be checked using a dial gauge and magnetic stand. The camshaft is installed in the engine and the lobes checked through the lifter bores. This can be done with the dial gauge stylus directly contacting the lobe or, if the dial indicator stylus is not long enough, on to the top of a lifter. If a lifter is used for the test it must be held in contact with the camshaft lobe using a small screwdriver, while the camshaft is rotated from lifter rise to lifter rise point across the heel or base circle of the camshaft lobe.

If the camshaft is located in the cylinder head, the camshaft is installed in the head and each lobe's base circle checked with a dial gauge and magnetic stand. If the head is cast iron, the magnetic stand can almost always be attached easily but, if the head is aluminium, a steel plate will have to be bolted to the cylinder head and the magnetic stand fixed to that. In all cases, it takes five minutes to run a dial gauge stylus over the base circle of a camshaft's lobes. The time to find any error is now; **not** after the engine rebuild is complete.

Note that base circle run-out is not usually a consideration with mechanical cams as they only lift the valve when the valve clearance is taken up.

Chapter 4
Camshaft Timing Principles

Most high performance camshaft manufacturers and re-grinding companies advise their customers to use a timing disc (for all intents and purposes a 360 degree protractor) and usually supply one to check the timing of the camshaft. The timing disc is bolted to the crankshaft, a temporary pointer fitted and the disc set to true TDC: the degrees of full lift for the inlet valve of the number one cylinder are then read off. The camshaft is either retarded or advanced (via vernier sprockets or offset keys) until it is timed to specification. Some camshafts are timed at particular lifter rise and lifter fall points. In both cases the disc is fitted to the engine, the camshaft timed and the disc is removed.

It's all very well to use a timing disc and bolt it on whenever the camshaft timing is going to be checked but, in reality, this is not always practicable. It's a fine method when the engine is being built up from scratch for the first time, but not necessarily after the engine has been run for a bit and you suddenly find that engine power has dropped off.

Then, to eliminate the camshaft from the possible problem list, you'll want to check the camshaft timing to see if it is still correct. With some engine installations (and that's most of them) it can be difficult to set up a timing disc easily. It's also fair to say that, with the engine installed in the car, it may well be difficult to see the crankshaft damper/pulley markings but, at least if the camshaft timing marks are accurately positioned on the damper/pulley, it will reduce the total amount of work in checking camshaft timing for error.

Accurate crankshaft pulley/damper markings are also useful if you want to experiment with camshaft timing during performance testing. If the camshaft timing is being altered in an effort to find the best overall setting it can be quite time-consuming setting up the disc for each adjustment.

The recommendation is to permanently mark the crankshaft damper or pulley with the relevant camshaft timing events (full lift point of the inlet valve of number one cylinder and, perhaps, the inlet opening and closing points). This means that the crankshaft damper/pulley will have a TDC line on it, plus the usual ignition timing marks as well as the full lift timing point of the camshaft at the very least and, if the engine has more than one camshaft, more than one full lift timing point or an alternative datum point (such as inlet closing point or exhaust opening point).

Once the pulley/damper has been accurately marked, all that has to be done to check camshaft timing is to remove the camshaft cover or rocker cover, fit a dial gauge to the valve retainer or follower of the inlet valve of number one cylinder, rotate the engine to the full lift timing position and see if the timing has moved from the original setting. This is very relevant to chain-driven camshafts: the chains wear and camshaft timing can become retarded. This form of check can usually be accomplished in ten to fifteen minutes and, further to this, if the camshaft timing needs adjusting and the engine is equipped with an easy form of adjustment (vernier sprockets/offset dowels) that, too, can

SPEEDPRO SERIES

be altered relatively easily.

Unfortunately, with some engine designs there is no way out of doing a considerable amount of work to adjust camshaft timing. If such an engine's crankshaft damper/pulley is permanently marked, it will be reasonably easy to check the timing, but adjusting it will still be difficult. Such an engine may need the timing chain cover removed (to get at the camshaft sprocket) and that may well mean removing the water pump (and everything associated with that), oil pump, distributor and disturbing the timing chain cover to sump (oil pan) seal. With the timing chain cover removed the camshaft timing can be adjusted by whatever means and then checked by setting up a timing disc and pointer, or by putting the timing cover back temporarily and using the permanent marks and the TDC pointer as per normal. What this means is that a combination of the two systems is used to set and check the camshaft timing.

Even if there is considerable work to effect a camshaft timing change or to bring the camshaft to the manufacturer's specified setting, it might produce a worthwhile gain in power. There is only one way to find out and that is to try it. On some engines it is going to take an hour or two and on others ten minutes. Most camshafts work best timed to their manufacturer's specifications, but this is not always the case. Slight retardation or advancement might bring about as subtle a change as improved mid-range power with no improvement whatsoever at the top end. The engine is, however, effectively better with the change, irrespective of how small the improvement is.

CHECKING TDC

The TDC mark of an engine is used as a datum point for so many things that it **must** be accurate. It doesn't actually have to be checked out as being dead accurate if a bolt-on degree wheel is going to be set-up in conjunction with true TDC by measuring the piston movement, but when the timing disc is going to be lined up with the engine's original equipment TDC markings, these markings have to be 'dead right.'

If the TDC position is not dead accurate and the camshaft timing marks done in relation to it, the camshaft timing is flawed from here on in. It is **imperative** that true TDC be accurately ascertained on any high performance engine.

Determining the true TDC point is best done with the cylinder head removed from the engine and then a dial gauge or a fixed stop is used to ascertain the precise point of TDC. Obviously, if the engine is assembled and it is not necessary to remove the cylinder head for any other reason, another method will have to be used (fortunately, there are several).

If a dial gauge is used, the crankshaft is rotated so that the piston crown comes up to the same point (just before and after TDC) in both directions and the crankshaft damper temporarily marked at these two points. True TDC is the mid-point

Dead stop comprises flat bar, adjustable screw and locking nut.

between the two temporary marks you have made. Do not be tempted to try to read the TDC point with the dial gauge: in effect, the piston 'stops' for several crankshaft degrees at TDC. The dead stop method is described in the following paragraphs.

If the engine is fully assembled, a common method of finding/checking true TDC is to use a dial gauge and stand with a long stylus through the sparkplug hole and contacting the top of the piston. Alternatively, core solder can be inserted through the sparkplug hole and positioned between the top of the piston and the flat part of the cylinder head gasket surface and used as a 'dead stop.' With the latter method, the crankshaft is rotated clockwise until the piston crown contacts the solder, stopping further rotation; the position of the TDC pointer on the block in relation to the crankshaft damper/pulley is then temporarily marked onto the damper/pulley. Next, the crankshaft is rotated anti-clockwise until the piston stops again; the crankshaft damper/pulley is marked again at this point. True TDC is at the exact mid-point between the two lines you've just made on the pulley/damper. Obviously, with this method, care must be taken not to squash the core solder or allow it to move and the core solder must be thick enough to act as a dead stop, but, all in all, this is a very effective and quick way of ascertaining the true TDC point of an engine if it has a 'squish' area. This method is more for side fitted sparkplugs where vertical access to the top of the piston for the dial gauge stylus is difficult, if not impossible.

In the worst case scenario, where the piston has an odd shape or contour on the top of it, and the sparkplug is side mounted, a sparkplug will have to be modified to form a

CAMSHAFT TIMING PRINCIPLES

Number one cylinder at TDC. The dial indicator is registering the highest point of piston travel. Pointer in line with TDC groove in crankshaft pulley.

Dial indicator at the prescribed point BTDC. Pointer is to the right of the TDC groove in the crankshaft pulley. Mark the crankshaft pulley in line with the fixed pointer.

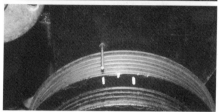

Dial indicator needle at the prescribed point ATDC. Pointer is to the left of the TDC groove in the crankshaft pulley. Mark the crankshaft pulley in line with the fixed pointer. True TDC is at the midpoint between the two temporary markings - in this case, the manufacturer's TDC mark was accurate.

dead stop. This means breaking up a sparkplug to get the threaded portion and brazing a short length of 0.312in/8mm diameter mild steel bar into the end of it, so that the piece of bar protrudes into the combustion chamber enough to contact the top of the piston a little way **before** TDC. The end of the protruding steel bar should be radiused to avoid marking the top of the piston. Turn the crankshaft gently to the point of piston contact in both directions. **Caution!** - Check that the valves are not able to contact the piece of round bar brazed into the end of the sparkplug as the engine is rotated, otherwise the valves could be bent.

Do not ever take it for granted that the standard marks machined in by the manufacturer of the engine are dead right. Check to see that they are right and, if found to be wrong, alter the position of the marks on the crankshaft damper/pulley or move the block mounted TDC pointer so that TDC is accurately registered.

MARKING THE CRANKSHAFT DAMPER/PULLEY

To mark the damper or pulley dead accurately it must be removed from the engine. You'll need a white piece of paper, or cardboard, about 12x12in/30x30cm, a 360 degree protractor, a pair of dividers, a 12in/30cm ruler, a small 90 degree set square, a fine felt tipped pen and a pencil or ball point pen.

Mark the centre of the paper and then measure the diameter of the damper/pulley. Using the dividers set to the radius of the damper pulley, draw a circle on the paper from the centre point.

Next, using the 12in/30cm rule draw a line through the centre of the circle extending 1in/25mm, or so, beyond the circle. Mark the top of the line 'TDC' (top dead centre) and the bottom of the line 'BDC' (bottom dead centre).

Note that the damper or pulley is going to be placed face down on to

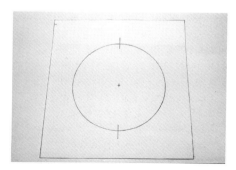

Paper with centre mark and circle of same diameter as pulley/damper.

SPEEDPRO SERIES

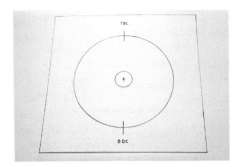

Straight line added to show TDC and BDC.

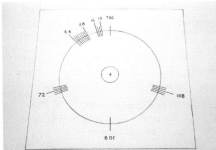

All relevant degree markings accurately marked around circumferance of the circle.

Damper face down on paper, ready to have timing points transferred.

Damper in milling machine having new degree marks machined in.

the paper, which will mean that, from now on, all degree markings on the paper are going to be in reverse. However, the figures will transpose from the paper to the damper correctly.

Place the 360 degree protractor exactly in the centre of the circle with the zero degree point on the TDC line. The degree markings that are applicable to the particular engine can now be placed on the paper.

The camshaft timing shown in the illustrations is 284 degree duration with 32 - 72 - 73 - 31 phasing and a full lift point of 108 degrees ATDC. The example shown includes all degree markings (including ignition) that this engine is ever going to need. Extra lines in two degree increments have also been placed on the damper so that accurate readings can be taken from either side of the prescribed timing points.

Moving anti-clockwise from the TDC mark in the example, you'll see 10, 12, 14 and 16 degrees for the ignition timing. 28, 30, 32, 34, 36, 38, 40, 42 and 44 degrees have also been marked. The 28, 30, 32, 34 and 36 degree marks are for the camshaft timing, while the 36, 38, 40, 42 and 44 degree marks are for the ignition timing. The 36 degree point doubles up here as it could, in fact, be used for ignition timing or camshaft timing. Also marked is the 72 degree point and 4 degrees each side of it in two degree increments. 104, 106, 108, 110 and 112 degree points are marked, too.

Note that cam manufacturers/grinders use various lifter rise points or checking heights at the opening and closing points of their camshafts. Some grinders use 0.004in/010mm, 0.010in/0.25mm, 0.016in/0.40mm, 0.020in/0.50mm, 0.025in/0.060mm, 0.050in/1.25mm or 0.053in/1.325mm, plus a few other heights as well. This doesn't matter, provided you know what the height is and what the relevant degrees are. Even after all this information is placed on the damper or pulley, the bottom line is that the camshaft should be timed to where the engine runs best. The placing of these prescribed degree marks is the best possible guide you can have but, ultimately, the camshaft may, or may not, end up being timed exactly as specified.

The damper/pulley can now be placed face down on the piece of paper so that its TDC mark is exactly in line with the TDC line drawn on the paper, and it is concentric with the diameter circle.

At this point, all the degree points on the paper are in true alignment with the damper/pulley and can be transferred using a 90 degree set square and a fine felt-tipped pen. After all of the points have been marked onto the damper/pulley, double-check that the original TDC mark on the damper is in line with the TDC line of the drawing, and that the damper is still concentric with the drawing diameter. Then check all of the degree markings to make sure they are all perfectly in line with the drawing's timing points.

Note that, as true TDC has already been verified, the extra marks added to the damper/pulley can be relied upon to be accurate.

CAMSHAFT TIMING PRINCIPLES

Filing degree marks into pulley/damper rim.

The damper can now be clamped in a vice on a milling machine table to have each timing point machined permanently across its outer rim. Alternatively, the lines can be very carefully filed into the damper/pulley using a triangular needle file. Use an engineer's set square to check that the line being filed is at 90 degrees to the front face of the damper/pulley. File the line lightly at first, check it to see that it is indeed square and then file the line to a depth of 0.025in/0.6mm, or so.

Appropriate numbers/letters are then stamped onto the surface of the damper/pulley to identify the new marks so that there is no confusion now, or later, about what each mark is actually for. For additional clarity, white paint can be worked into the stamped numbers and main degree marks to highlight them.

CAMSHAFT TIMING CHECK USING THE FULL LIFT POSITION

You can only get out of a high performance engine what you put into it: sometimes, though, you can get a lot less out of an engine than you put into it. The reason for getting a lot less can be the need for a relatively simple adjustment.

One such simple adjustment is advancing or retarding the camshaft timing from the true standard position to see if a power gain is possible. The first thing to do, however, is make sure that the replacement camshaft is timed precisely as the manufacturer or grinder specifies, and then test the engine. The next stage is to move the camshaft into a position 2 degrees and then 4 degrees advance of this position and re-test engine performance. Then retard the camshaft timing by 2 degrees and then 4 degrees from the original recommended setting and test the engine again to see which of the settings gave the best performance for your requirements.

If a timing disc is to be used it must firstly be set in relation to TDC of the number one piston. It is usual to set up the timing disc independently, and not rely on the TDC marking of the crankshaft damper or pulley in the interests of absolute accuracy. However, if the engine's crankshaft damper/crankshaft pulley has been accurately marked, as recommended, no timing disc need be used.

The full lift timing point as used by camshaft manufacturer/grinder relates to the inlet valve of number one cylinder. It is also quite possible, and recommended, that the actual opening and closing points of the inlet valve be measured and noted, if only as a double-check to see how accurate the camshaft is. In the first instance, a dial gauge is used to measure the full valve lift. The dial indicator is then set to zero at the full lift point.

The crankshaft is then turned anti-clockwise 30 degrees, or so (on a clockwise rotating engine), before the full lift position and then brought back to a point 0.010in/0.25mm before full lift. An actual degree reading is taken off the timing disc if fitted, or the 'full lift' point is marked on the crankshaft damper/pulley with a white marking pen (white for clarity).

The important technical point here is that the position of the dial indicator needle is telling you exactly where the valves are in relation to full lift. The second part of this vital procedure is designed to pinpoint the exact position of full lift (to within a degree, or less) by picking up points each side of the actual full lift position. That is a 'set point' before full lift with an opening valve and that same set point with a closing valve. The term 'set point' meaning the exact amount of valve movement before the full lift point and the exact same amount of valve movement after full lift.

Note that the following examples use 0.010in/0.25mm, but you can use either 0.005in/0.125mm or 0.010in/0.25mm; one measurement may prove to be more convenient than the other on the basis of distance about the central full lift camshaft timing point. The full lift position in the example is 108 degrees after top dead centre for the inlet valve (see diagram on page 32).

The crankshaft is then turned past full lift to when the 0.010in/0.25mm point on the dial gauge comes up again (see diagram on page 32). A reading is taken off the timing disc again, or the crankshaft pulley/damper marked. The true full lift position is in the middle of the two readings taken at the stopping points. This exact number of degrees is then related to the specification sheet as supplied by the camshaft manufacturer or regrinder.

Note that the crankshaft should be rotated the way it normally turns to avoid any slackness in the camshaft drive interfering with the readings.

The camshaft manufacturer's full lift timing position should always be used and tested in the first instance.

SPEEDPRO SERIES

Valve at full lift and the dial gauge needle is set on zero.

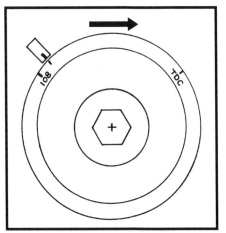

The dial gauge needle is 0.010in/0.025mm before the zero position. Mark the crankshaft pulley in line with the pointer.

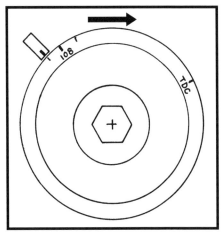

Dial test gauge needle is 0.010in/0.025mm before the zero position. Mark the crankshaft pulley in line with the pointer.

Then reset and tested at 2 degrees and 4 degrees advance and then 2 degrees and 4 crankshaft degrees retard. The engine may perform better in any of these positions. After testing the engine with various suggested settings, set the camshaft timing to give the best engine performance for your own requirements. Note that there is virtually never any point in advancing or retarding the camshaft by more than 4 degrees from the listed setting. If there is no difference between any of the settings suggested, run the engine at the camshaft manufacturer's specified setting.

The full lift timing mark, plus 4 degree advance and 4 degree retard marks (the two degree advance and retard points are quite easy to estimate) can be permanently marked onto the crankshaft damper/pulley for future use. Permanent marking of the crankshaft damper/pulley is strongly recommended as the marks are always there and setting-up of a timing disc is unnecessary in the future. Further to this, the camshaft timing can literally be checked anywhere (race meeting, for instance) provided you have a dial gauge and stand.

Some camshaft grinders recommend that their product be timed at a lifter rise and lifter fall point such as 0.020in/0.50mm. Broadly speaking, this is similar to the full lift timing method except that the points are much further away from full lift.

Some engines have generally accepted timing mark points, such as Harley-Davidson motorcycles, for example. They almost always use 0.053in/1.35mm as a lifter rise point for camshaft timing and 0.053in/1.35mm lifter fall point to check/time camshafts. Harley-Davidson, and nearly all aftermarket camshaft manufacturers who make camshafts for H-D engines, use the 0.053in/1.35mm lifter rise/fall points to rate their products. Everyone

CAMSHAFT TIMING PRINCIPLES

knows what they are getting in duration.

If a camshaft is recommended to be timed at, say, 0.020in/0.50mm lifter rise/lifter fall of the inlet valve of number one cylinder, a dial gauge is positioned to measure the valve lift (degrees BTDC) and, when the crankshaft is at the prescribed number of degrees, the required amount of lift must be present. If the crankshaft is turned on through full lift to when the valve is closing, the 0.020in/0.50mm point (degrees ABDC) will come up again on the dial gauge: again, the crankshaft must be at the prescribed number of degrees. If the camshaft timing is wrong, it must be adjusted. The crankshaft damper/pulley can be permanently marked with the specified degrees, or a degree wheel fitted to the crankshaft whenever necessary. The camshaft can then be advanced or retarded in relation to these settings.

It's actually easier to use the full lift position for camshaft timing on most engines (not all) as, overall, you're dealing with one point instead of two. Full lift timing is regarded as the simplest way of getting a good result as the timing is either correct or in front or behind the mark. Irrespective of how the camshaft is recommended to be timed, the bottom line is that the camshaft timing needs to be set for best all-round power production/torque and this may, or may not, be at the manufacturer's recommended setting.

To avoid any confusion, it is recommended to set the camshaft at the full lift position exactly as prescribed by the camshaft grinder/manufacturer and not bother with the actual opening and closing points at all. Part of the reason for not bothering too much about the opening and closing points of the camshaft is that nothing can be done to them anyway (except regrinding).

The problem that full lift timing can pose is that the exact opening and closing points are not necessarily known and, therefore, the actual duration (specific number of degrees) of the camshaft could be less than the design's specification. The point being that camshafts, as ground, do not always have the right number of duration degrees in total. A camshaft lobe can have up to 10 degrees of duration less than listed. Seldom do camshafts have more than advertised duration.

All too often an engine has a high performance camshaft fitted and it is automatically assumed that the camshaft is timed right. If you're lucky the camshaft may be timed right, but the chances are that it will not be. The reason for this is production tolerances and, perhaps, cylinder head planing (retards the camshaft timing on OHC engines).

What can happen if the camshaft timing is, say, well retarded (minus 3-8 degrees, for example) is that engine response will be less than it could be. As it happens, not all engines are camshaft timing sensitive: some can be run up and tested timed at 3 degrees retarded, then at the recommended setting and then advanced by 3 degrees without perceptible difference in engine response or power between any of them. Some engines respond to being set-up and run 3 or 4 degrees retarded for high rpm applications, and 3 or 4 degrees advanced for low to mid-range applications. Such engines always run well at all the settings, it's just that they run a little better at one setting than the others and the only way to find the optimum setting is to try various settings. Such experimentation can mean quite a lot of work and, in the end, it could all be for nothing; conversely, the engine could be livened-up considerably. One thing is for sure, if an engine has a camshaft installed with no thought given to how well it is actually timed, you'll never know if you're getting optimum power. No claim is made that engine power will increase by moving the camshaft timing around, just that it might improve.

As a general rule, engines which have the inlet and exhaust lobes on one camshaft have the camshaft set at the inlet valve's full lift point of the number one cylinder. Engines which have the inlet lobes on one camshaft and the exhaust lobes on a separate camshaft are best adjusted at the inlet closing point and the exhaust valve opening point (0.001in/0.025mm valve movement from the fully seated position).

Visit Veloce on the Web - www.veloce.co.uk

Chapter 5
Camshaft Problems

ENGINEERING TOLERANCES

All engineering components are made to specifications and, as such, have manufacturing tolerances (or allowances) on those specified sizes. Manufacturers of engineering components make the components as accurately as possible, but there is, one way or another, an allowance for error - however slight that error may be. In the overall scheme of things, the error in components is virtually always very, very small, to the point that it can be insignificant and, basically, not matter. The problem is that the camshaft timing on a performance engine does matter and, in many instances, engine power is lost through component error (not intentional or entirely preventable).

The possibilities for error include the following. The crankshaft keyway can be slightly out either way, the crankshaft drive sprocket keyway slot can be out, the timing chain can be loose, the camshaft drive sprocket's keyway slot can be out, the keyway in the camshaft can be out and the camshaft lobe phasing in relation to the keyway can be out. Any one of these things could be slightly out, with some plus factors cancelling out some minus factors, but, almost always, the camshaft timing will end up retarded with the largest error caused by the timing chain. It is just as possible that all tolerance factors lead to the camshaft timing being dead right, but you will never know unless you check. What is being advocated here is that the possibility of error, irrespective of how it has occurred, should be taken into account and removed from the equation.

CAMSHAFT SPECIFICATIONS

There has always been some confusion with regard to camshaft specifications, camshaft timing and how, in fact, to time a camshaft correctly for optimum results. It is very easy to get lost with camshaft timing because there are so many variables with camshafts and, to be more specific, with camshaft lobes. There are various methods of timing camshafts and ideal methods of timing the various types of camshaft. All of the methods have been developed over the years to remove the problem associated with the difficulty of grinding a camshaft lobe that is completely accurate in relation to the design specifications.

The fact is that most camshaft lobes, as ground, are not completely accurate, but they are usually accurate enough, except in extreme circumstances. When a camshaft is chosen from a catalogue, for instance, it is fair to assume that the camshaft will meet the design specification (such as degrees of duration and lift). However, not all camshaft lobes actually make their design specifications (especially at very low lift) and this factor can cause all sorts of problems when timing a camshaft (depending on how the camshaft is timed and how it is possible to time the type of camshaft). All of this leads back to the full lift point as it is reasonably easy to locate.

If a camshaft with poorly ground lobes is timed at the full lift position, and then the absolute inlet opening

CAMSHAFT PROBLEMS

and closing points checked, each opening and closing point could be anything from 2-5 degrees different from what is claimed by the camshaft manufacturer. It's fair to say that camshaft manufacturers/grinders list duration figures at certain checking heights.

If a camshaft is timed at the inlet opening and closing points of number one cylinder, for instance (as some people do), the camshaft could be out of time and engine performance will, possibly, not be as good as if the camshaft were timed at the true full lift position. This is because of the difficulty of accurately picking up the actual opening and closing points. Admittedly, many camshafts will be right at the full lift timing point and the opening and closing points of the inlet and exhaust valves - but far too many are not.

Take, for example, a camshaft listed as having 42-78-78-42 phasing with full lift inlet timing at 108 degrees ATDC and 0.450in/11.4mm of valve lift. If such a camshaft is set up in an engine at the advertised 108 full lift inlet timing, the opening and closing points could actually check out to be 40-75-77-41 for the number one cylinder, and the actual valve lift could be anything from 0.420-0.435in/10.7-11.0mm. Seldom will a camshaft lobe give the full advertised amount of lift. This is not to say that all of this 'error' matters too much in the overall scheme of things, but it certainly doesn't make it right. It can get confusing (not to mention annoying) to the point that you could wonder whether you have the right camshaft in some instances!

One thing is for sure, an engine equipped with a camshaft accurate to the design specifications goes better than the same engine equipped with a camshaft which is incorrectly ground.

LOBE PHASING

When a camshaft is timed we do not set-up on each lobe (although it is possible to check every lobe of a camshaft when it is installed in the engine). Instead, for convenience, we use the inlet lobe and, perhaps, the exhaust lobe of the number one cylinder to time the camshaft. This means that, in the overall scheme of things, we are assuming that the camshaft grinder has done everything correctly and, 99% of the time, this is so.

If a camshaft is suspected of being inaccurate, the procedure for checking it is to set up each cylinder in turn at TDC (using a timing disc) and read off the opening and closing points and the full lift timing points of all the inlet and exhaust valves. This involves some work, but it's the only way to check the phasing of all the lobes of a camshaft and it does give a true picture of the integrity of the camshaft.

This problem of faulty lobe phasing does crop up occasionally, but it is not a frequent occurrence. In short this means that if the camshaft has been ground incorrectly, the inlet of the number one cylinder could be correctly set to the camshaft grinder specifications of, say, inlet opens 42 degrees BTDC and closes 78 degrees ABDC but, because the camshaft was ground incorrectly, the inlets of the other cylinders could be phased anywhere between plus or minus 5 degrees, or more. A camshaft could be phased so that the inlet opens 37 degree BTDC and closes 83 degrees ABDC, or opens 47 degrees BTDC and closes 73 degrees ABDC or, worse, with the other cylinders anywhere in between. The exhaust phasing could also be out.

Correct lobe phasing, lobe to lobe, is vital but is **always** the camshaft manufacturer's/grinder's responsibility: it is possible for them to get it wrong.

RE-GROUND CAMSHAFTS

Just because a high performance camshaft is a re-grind (and that is about 50-60% of them) does not mean that it is inferior to a new camshaft. Many small camshaft re-grinding businesses (one-man-bands) do perfect work and take absolute pride in the fact that any camshaft they re-grind is suitable for re-grinding to start with, and every lobe is phased to within 0.5 (30 minutes) of a degree, or less, and each lobe is perfectly ground for shape and size. Errors are very rare with this sort of business.

There is always going to be the odd time where a camshaft should not have had a particular grind put on it, such as one that undercuts the core diameter too much and weakens the camshaft. Camshafts like this may break in service and cause a lot of engine damage but, overall, failures are few and far between - provided everything associated with the valve train is set-up correctly. There is nothing wrong with a re-ground camshaft provided everything is done correctly and the camshaft being re-ground is suitable for re-grinding in the first place.

Complicating the choice of a suitable camshaft for re-grinding is the fact that the camshaft you have may have been re-ground before. It may well seem to be only slightly worn but, in fact, has been well ground down and is effectively unserviceable because it was re-ground previously when it should have been scrapped. The problem here is that the actual size specifications of camshaft lobes are not easy to get hold of and this means that it can be difficult to say whether a particular camshaft is a standard but worn camshaft with lobes that are basically on size, or a camshaft which has good lobes but has been ground down to clean up the camshaft lobes to make them serviceable again.

The solution to this problem is to look at a few camshafts from the same type of engine and gauge what the average lobe size is (that is standard but worn). In the overall scheme of things, it's unlikely that three camshafts, for instance, will be found from three different sources and all of them have been ground excessively. Most camshaft re-grinders know from experience the basic sizes of camshaft lobes and they can tell, by simply looking at the camshaft presented to them, whether or not it is suitable for re-grinding. This doesn't always work; many unsuitable camshafts have been re-ground in the past which has caused the whole debate about the quality of re-grinds to come up in the first place.

Nearly all camshafts can be re-ground, but only if all of the lobes are still basically 'on size.' Some wear is acceptable, but excessive wear is **not**. If one lobe on a camshaft is well down (0.040-0.060in/1.0-1.5mm) on its nose to base circle dimension compared to all other lobes on the same camshaft, the camshaft is not really suitable for re-grinding because of this one lobe. If all of the camshaft lobes are very similar in size (within 0.005-0.010in/0.13-0.25mm when measured from the base circle to the nose) then the camshaft will take a regrind quite successfully. The reason that such a camshaft will take a re-grind successfully is because the camshaft is worn overall but, nevertheless, in a good condition. A well-worn camshaft lobe indicates a problem: it could be as simple as that the particular lobe was never up to specified hardness (incorrect heat treatment) from day one, or something went wrong with the lifter which wore the camshaft lobe. It is very unusual for all of the camshaft lobes on a camshaft to be worn down to minus 0.060in/1.5mm or more from the original lobe size. The largest inlet lobe and the largest exhaust lobe on a camshaft are the standard against which all other lobes are measured in the first instance.

Once the camshaft is in the re-grinding machine, the first check made is to see if the camshaft is actually straight. If the camshaft is not straight, it must be straightened. There must be a minimum 'run-out,' journal to journal, of not more than 0.001in/0.025mm overall.

Some standard camshafts are not able to take high lift performance profiles because the lobe is physically very small in relation to the core diameter of the camshaft. For the camshaft lobes to take a performance grind, too much material would have to be removed from the base circle of the camshaft lobe, and this might mean that the core diameter of the camshaft gets severely undercut. This is not desirable as the camshaft is weakened, especially if a sharp corner is formed when the camshaft core is undercut.

Such camshafts could be built up using a hard surfacing agent, but this is costly and not normally considered. One or two lobes of a camshaft can be repaired to restore those individual lobes, but to build up all of the lobes is very expensive. That said, in some cases it is the best option, if not the only one.

A good camshaft to re-grind is a used high performance camshaft that has been ground on a 'blank' as made by one of the aftermarket camshaft manufacturers. These camshafts are frequently made with a slightly smaller core diameter (0.040in to 0.060in/1.0mm to 1.5mm) than an original equipment camshaft and a lobe base circle diameter slightly less (0.040in/1.0mm) than an original equipment camshaft. With the lift invariably being much more than the original equipment camshaft, very little, if any, material will have to come off the base circle of the camshaft lobe except for a 'clean-up' allowance of 0.002in/0.05mm, or so.

A used 'wild' camshaft will almost always take a less radical grind with ease. The same camshaft checking procedure still applies: the camshaft must be dead straight and have no excessively worn lobes (indicating a 'soft' lobe). If the camshaft has a damaged lobe it may still be used, but only if the damage cleans up during the re-grinding process.

Some modern twin overhead camshaft engines need to have any performance profile ground on to a 'blank' because the base circle diameter cannot be reduced from standard (well, not by much). The reason for this is that the followers are almost always hydraulic and the internal adjustment on the followers is quite limited. If the base circle is reduced too much the follower cannot compensate for it. Also, the followers are very short on some engines and, the more the lobe base circle is ground down, the higher the follower lifts out of the bore in the cylinder head (high bore wear can be the result). There is also the prospect of follower contact with other parts of the camshaft if the base circle is reduced too much.

If hydraulic camshafts are changed to mechanical action, things are not so bad and most replacement mechanical followers are longer and have an infinite adjustment, therefore

CAMSHAFT PROBLEMS

the lobe can be ground to almost any size without detriment.

LACK OF POWER (CAMSHAFT AT FAULT?)

When modified high performance engines do not perform as well as expected there can be many reasons and, although the camshaft is often blamed, in reality, it is seldom the cause of the problem. The camshaft, in most instances, will be doing precisely what it was designed to do, but that does not mean that the rest of the engine is up to the same standard.

If the camshaft is timed correctly and all of the lobes proven to be phased correctly, then the problem does not rest with the camshaft. Tight valve clearances (or a near lack of valve clearance) can cause problems but, that aside, any problems associated with lack of power are elsewhere.

It is not easy to check the camshaft timing and even worse to adjust the camshaft timing on many engines. Such engines require the timing cover to be removed, which means the oil pump is removed, the distributor is removed and the sump (oil pan) seal to the timing cover is disturbed. External markings on the crankshaft pulley/damper are essential to set up an engine's camshaft timing and then be able to check the camshaft timing relatively quickly. The time and effort taken to make these marks is well worthwhile.

Most engines that get modified are standard production units and the manufacturers have designed these engines to power conventional road going cars and built them accordingly. When it's decided to modify one of these engines criteria change and power rather than fuel economy, smooth running and low speed torque becomes the goal. Fitting a high performance camshaft with 280 degrees, or more, of duration into an otherwise standard engine is a waste of time. Very few engines respond well to a straight camshaft change. What engines do respond to is a whole raft of changes which improve volumetric efficiency, such as a modified cylinder head, better carburation, better exhaust system and a better ignition system to ignite the charge at the correct time in the cycle. The actual point of ignition (degrees BTDC) that the ignition is actually fired at must be one of the least considered aspects of many high performance engines, but one that causes a huge power loss if not right! *(see "How to Build & Power Tune Distributor-Type Ignition Systems" by Des Hammill from Veloce Publishing).*

One problem that causes an engine to end up 'over cammed' is the fact that all high performance camshafts cost a similar amount, so the mistaken idea of getting maximum duration for the money seems to prevail. The rule is **'never buy more duration than is really needed.'**

Chapter 6
Timing Procedure - Cam-in-Block Engines

The usual run-of-the-mill camshaft-in-block production engine has either chain-driven or gear-driven camshafts, the majority being chain-driven. These engines have only one camshaft and all of the lobes (inlet and exhaust) are ground on to the same camshaft and phased in relation to each other during the grinding process. The only part of the camshaft timing that is adjustable is the position of the camshaft in relation to the crankshaft. This means that the camshaft can be either run to the camshaft manufacturer's recommended setting, or advanced or retarded from this setting. Normally, adjustment of camshaft timing is by an offset key in the camshaft drive sprocket/gear, an adjustable (vernier) camshaft drive sprocket/gear or a multi-keywayed sprocket/gear.

The camshaft in block design of these engines means that, to alter the camshaft timing, there is going to have to almost always be some dismantling of the front of the engine. Aftermarket manufacturers of engine components have made timing chain covers (for some of the more popular engines) that feature removable plates which allow easy access to the camshaft sprocket and, more specifically, to the securing bolts of an adjustable camshaft sprocket system. On the other hand, builders of modified engines often modify the original equipment parts so that a plate can be removed and the timing adjusted in a few minutes. The camshaft sprocket used is a modified original one, or an aftermarket adjustable one.

Original equipment timing chain cover which has been modified for easy access to the camshaft sprocket.

TIMING PROCEDURE CAM-IN-BLOCK ENGINES

If easy access to the camshaft sprocket can be arranged it is definitely a good idea, even if only to re-adjust the camshaft timing back to the specified setting (for example, once the timing chain has worn a bit and retarded the camshaft timing). Timing chains do wear, some of them quite quickly. If your particular engine has a chain tensioner, making a reasonably simple adjustment to camshaft timing can restore correct timing. Engines that do not have chain tensioners end up with fluctuating camshaft timing (and ignition timing, if camshaft driven) and the only solution is to replace the timing chain. Always fit the best possible quality chain to any high performance engine and, if the original chain is a single row (simplex) type, replace it, if possible, with a duplex chain and sprockets from an aftermarket source, or original factory equipment from the sports version of the same engine, if available.

On most American V8 engines, for example, there is a considerable amount of work to be done to get at the timing chain. The water pump may well be an integral part of the timing chain cover, the distributor may bolt on to the timing chain cover and, further to this, the oil pump may well be fitted onto the timing chain cover. The whole lot will have to be removed to effect a simple adjustment (Buick/Oldsmobile/Rover V8). Not all American V8s are like this.

The small block Chevrolet V8 engine, for instance, only needs to have the water pump removed but, still, removal of the timing chain cover is not a straightforward job because the base is fitted into the oil pan (the timing chain cover is fitted before the oil pan). However, the timing chain cover can be modified and fitted with a removable plate to make this job easier.

If the crankshaft damper/pulley is permanently marked with full lift degrees, the camshaft timing can be more easily checked and the chain wear monitored. Once the camshaft timing has been set for optimum engine performance, the marks are used to check the rate of wear of the components and re-adjust the camshaft timing when necessary.

In view of the necessity to remove quite a lot of componentry from the front of one of these engines, the ideal time to set the camshaft timing to the camshaft manufacturer's setting is when the engine is being assembled/rebuilt. In this situation, the camshaft lift is directly measured for full lift at the top of the lifter/follower, using a dial gauge and stand in conjunction with a timing disc mounted on to the crankshaft, and a makeshift pointer.

CHECKING FOR FULL LIFT USING THE LIFTER

Three things are essential to check or set camshaft timing. The first is the specification of the camshaft concerned, the second is true top dead centre (TDC) of the piston of the number one cylinder and the third is the true position of the full lift of the inlet lobe for the number one cylinder. Make sure that all of the details about the camshaft are known before the camshaft is bought.

FINDING 'TRUE' TOP DEAD CENTRE

Any engine, when being assembled, can, and should, be checked to see how accurate the top dead centre (TDC) mark is. This involves relating the piston crown's position at top dead centre to the crankshaft damper/pulley. There are two ways of doing this easily on a partially assembled engine. The first involves the use of a 'dead stop' which is a bar fixed to the top of the block with an adjustable screw in a position so as to contact the top of the piston.

Modified timing chain cover with the sprocket cover removed. This sprocket is slotted to allow only timing advance (about 8 degrees).

SPEEDPRO SERIES

Dead stop bar bolted onto the top of a block. The adjustable screw (with locking nut) in the middle is in contact with the piston crown.

Crankshaft stopped 8 degrees after top dead centre (ATDC) - the crankshaft has been turned anti-clockwise until the piston contacts the stop.

Crankshaft stopped 8 degrees before top dead centre (BTDC) - the crankshaft has been turned clockwise until the piston contacts the stop.

Crankshaft at true top dead centre (TDC) as confirmed by using the dead stop.

The dead stop's screw is adjusted and locked so that the piston is stopped just before it gets to TDC, irrespective of which way the crankshaft is turned. The piston stops at exactly the same distance up the bore irrespective of which way the crankshaft has been turned. Make a temporary mark on the crankshaft pulley/damper to coincide with the piston's stopping point in both directions of crankshaft rotation.

The centre position between the two marks is the 'true' TDC of the piston. In this case, shown in the photos, the marks are correct and the TDC line is spot-on. The block's pointer can be altered if it is in the wrong position. In the example shown, the pointer is relatively easy to reposition. Admittedly, not all pointers are easy to move and, in these cases, the damper/pulley rim can easily have a new mark added in the correct position.

The second method of finding true TDC involves the use of a dial gauge and a magnetic stand or a clamp arrangement. The end result is the same as using a dead stop, but the principle involved in achieving the required result is slightly different. The common factor, though, is still to ascertain two checking points. Both methods are accurate, but the dead stop method is totally foolproof.

With the dial gauge positioned directly above the piston crown of the number one cylinder, the piston is brought to what appears to be TDC and the dial of the gauge zeroed. The maximum needle movement is the highest point of piston travel.

The problem with ascertaining TDC is that the piston effectively stops, or 'dwells,' at TDC as the crankshaft approaches, gets to and goes passed TDC. The longer the stroke of the crankshaft, and the longer the centre-to-centre distance is of the connecting rod eyes, the more pronounced this factor is. The solution is to measure the rising piston crown at a set point and the descending piston crown at the same point down the bore.

With the piston crown approaching (ascending) TDC, stop rotating the crankshaft when the dial indicator needle shows, say, 0.030in/ 0.75mm BTDC. Note that although 0.030in/0.75mm is used in this example, it is quite possible to use as little as 0.005in/0.125mm. If the dimension chosen is too large, the distance on the crankshaft damper/ pulley rim between the added marks will become excessive and accuracy could be reduced.

At this point the rim of the damper/pulley adjacent to the tip of the timing pointer is temporarily marked with a marking pen or, if there are sufficient degree markings on the damper/pulley, the relevant reading is noted.

The crankshaft is now turned past the TDC point until, with the piston now descending, it is stopped when the dial gauge needle once again registers 0.030in/0.75mm.

At this point the crankshaft damper/pulley is marked again using a marking pen.

The 'true' TDC is that point exactly midway between the two marks.

CHECKING FULL LIFT VIA LIFTER MOVEMENT

With a lifter, or a suitable mandrel, installed in the block on the inlet lobe

TIMING PROCEDURE CAM-IN-BLOCK ENGINES

Dial gauge with magnetic mount being used to check top dead centre (TDC). The dial has been set to zero.

Dial gauge needle registering 0.030in/0.75mm before the zero position (before top dead centre).

Dial gauge needle registering 0.030in/0.75mm of piston descent (below top dead centre).

Dial gauge clamped to an alloy block.

The damper/pulley will be somewhere before top dead centre (BTDC): mark the pointer position with a marking pen.

The crankshaft damper/pulley is now after top dead centre (ATDC); mark the pointer position.

of the number one cylinder, a dial gauge can be positioned to measure the full lift of the lobe as the crankshaft is turned. When the crankshaft is turned, the lifter will rise and, when at full lift, the needle of the dial gauge will stop moving. If the crankshaft is turned further the needle of the dial gauge will fall as the lifter drops back down. For this test no oil should be applied to the lifter so that it moves freely in the lifter bore. If the lifter does not move freely, light pressure may have to be applied to it when the lifter has gone past the full lift point to ensure that the lifter follows the camshaft lobe profile.

Turn the engine in its normal direction of rotation only and, if the engine has a timing chain tensioner fitted to it normally, make sure that it is fitted now. With the full amount of camshaft lift, read off the dial gauge, set the dial to zero and turn the engine over again. As the needle approaches the full lift point, go quite slowly and make sure that the needle is set to the true zero point as it reaches full lift and passes over it. Adjust the needle if there is any discrepancy so that zero is the highest point of needle travel reached.

Rotate the crankshaft again until the zero position is approaching again and, this time, stop turning the crankshaft 0.030in/0.75mm before the zero point is reached.

Read off the number of

SPEEDPRO SERIES

Dial gauge stylus positioned to contact the top of the lifter.

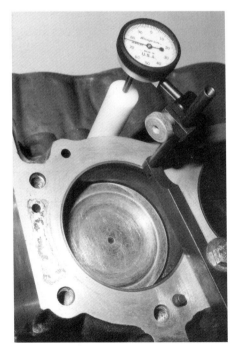

Dial gauge stylus positioned on a specially made mandrel.

A dial gauge positioned on a cast iron block with a specially made, close-fitting pin slotted through the pushrod hole to contact the follower.

Dial gauge registering 0.030in/0.75mm before full lift.

Pointer indicates a position before the full lift timing mark of 108 degrees.

Dial test indicator is registering 0.030in/0.75mm after the full lift position.

Pointer indicates a position after the full lift timing point.

crankshaft degrees present (only possible if the damper/pulley is fully degree marked), before the full lift degree mark on the crankshaft damper/pulley reaches the pointer. If there is only the specific degree point marked on the crankshaft damper/pulley, mark the damper/pulley at the 0.030in/0.75mm BTDC point. If the full lift point of the number one cylinder's inlet valve timing is, for example, 108 degrees ATDC, the pointer will be to the right of the 108 degree on the damper/pulley on a clockwise rotating engine.

Rotate the crankshaft further until the zero position is reached and slowly continue on rotating the crankshaft until the 0.030in/0.75mm point is reached again and then stop.

Mark the crankshaft damper/pulley adjacent to TDC pointer, or read off the degrees on the damper at this point.

At this stage the camshaft will have been stopped 0.030in/0.75mm before the full lift point, been through the full lift point and finally stopped at the 0.030in/0.75mm point on the other side of full lift. That point between the two 0.030in/0.75mm stopping points in crankshaft degrees of rotation is the true full lift position of the camshaft lobe.

TIMING PROCEDURE CAM-IN-BLOCK ENGINES

Be aware of the fact that the lifter may not automatically follow the camshaft lobe and drop as the nose of the camshaft goes past the full lift point. Light pressure applied with a small screwdriver may be necessary to ensure that the lifter does follow the lobe profile.

If the camshaft is not timed correctly it will have to be advanced or retarded until it is. With chain-driven camshafts it often pays to advance the camshaft timing 1 degree. The reason for this is that, when the camshaft is being turned by hand, there is no pre-load on the valve train as there would be when the valve springs are exerting resistance. Advancing of the camshaft timing at this point does tend to cancel out this factor. If the camshaft timing is retarded 1 degree, for instance, there is no way there is ever going to be less than 2 degrees of retardation when the engine is running with all the normal stresses on the valve train. This is a part of the reason why it's worth experimenting with small changes to camshaft timing once the engine is running - all of the variables are compensated for by finding which setting actually gives the best results.

CHECKING FULL LIFT VIA ROCKER ARM/VALVE RETAINER MOVEMENT

In the first instance, a dial gauge is positioned so that its stylus is on the valve retainer of the number one cylinder's inlet valve. The engine is rotated until the valve is at full lift, which is ascertained by noting when the needle stops moving. At this point the dial of the dial gauge is zeroed. Check to make sure that the dial gauge has enough travel to read the full valve lift. The 'true' full lift position is not the instant that the needle stops, because camshaft lobes have an amount of 'dwell' at full lift. The 'true' full lift is that point in the middle of 'dwell.'

Note that most engines rotate clockwise - but not all. Any engine being checked must be turned in the direction that it normally turns in.

In the example that follows, the crankshaft pulley has been marked at 108 degrees for full lift which suits the particular camshaft fitted to this engine. The inlet full lift position of nearly all camshafts ranges between 100 and 115 degrees ATDC. In the example shown in the photographs, the true TDC point of crankshaft pulley has been checked in relation to the TDC pointer and the 108 degree full lift point was marked (with a white marking pen) on the pulley using a 360 degree protractor. All of the crankshaft pulley marking was done without removing anything from the engine. All in all, it took about ten minutes to mark the crankshaft pulley and begin checking the camshaft timing.

With the full valve lift point known, the crankshaft is turned backwards a good 30 degrees before the full valve lift point and then brought back to a point 0.010in/0.25mm (0.005in/0.125mm can be used instead) before the full lift position. At this precise point stop turning the crankshaft and look at the crankshaft damper/pulley. Use a fine tipped white marking pen to mark the crankshaft damper/pulley adjacent to the pointer (or if a timing disc is fitted, read off the number of degrees).

The crankshaft is now turned further to beyond the full lift point until the 0.010in/0.25mm point is reached again and the crankshaft is then stopped. The crankshaft damper/pulley is marked again adjacent to the pointer (or if a timing disc is fitted, read off the number of degrees). That point exactly midway between the two

When the camshaft is timed correctly the damper/pulley will be here (108 degrees in this case). Inlet valve at full lift and the dial gauge needle set at zero.

marks (or the recorded degrees from the timing wheel) is the 'true' full lift position.

It is not difficult to see if the 108 degree (in this example) full lift point is in the middle of the two dots placed on the crankshaft pulley. Admittedly, no actual number of degrees are used to ascertain the precise point of full lift,

43

SPEEDPRO SERIES

Valve is positioned 0.010in/0.25mm before full lift. The crankshaft damper/pulley pointer is ahead of the 108 degree full lift point on this clockwise rotating engine.

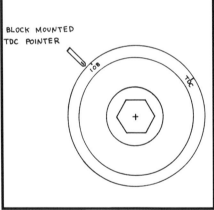

Valve is positioned 0.010in/0.25mm after full lift. The crankshaft damper/pulley pointer is after the 108 degree full lift point on this clockwise rotating engine.

but this procedure usually sees the camshaft timing accurate to within 1 degree. The alternatives to this method are to fit a degree wheel temporarily to the front of the crankshaft and rig up a pointer, or to mark the crankshaft damper/pulley with plenty of degree markings in two degree increments. The latter would mean having the damper/pulley marked to suit a 108 degree (in this example) full lift camshaft at 102, 104, 106, 108, 110, 112 and 114 degrees. The preferred method is to have the crankshaft damper/pulley accurately marked (once true TDC has been established) when off the engine. This way the degrees are directly read off the damper/pulley. This work must be done accurately, but it only has to be done once. With all the necessary marks on the damper/pulley the engine carries them around with it, facilitating checking/adjustment virtually anywhere.

Of course, it may prove necessary to advance or retard the camshaft timing to get the precise full lift position to coincide with the camshaft manufacturer's recommended camshaft timing.

Visit Veloce on the Web - www.veloce.co.uk

Chapter 7
Camshaft Timing Procedure - S.O.H.C. Engines

These engines have all of the lobes (inlet and exhaust) on the same camshaft, with the only possible adjustment being to advance or retard the camshaft in relation to the crankshaft. The form of adjustment, irrespective of whether the camshaft drive is by chain, toothed belt or gears, can typically be offset key, offset dowel, standard dowel in an elongated slot or an aftermarket adjustable vernier type sprocket. This is all much the same as for a cam-in-block engine, except that the chain or belt is much longer due to the fact that the camshaft is on top of the engine.

Single overhead camshaft engines can have the camshaft lobes virtually direct-acting on the valves via 'inverted bucket'-type followers, or through finger followers. It doesn't matter what type of actuation is used as long as a dial gauge stylus can be positioned on the valve retainer of the inlet valve of the number one cylinder.

There are several things that can cause an overhead camshaft engine to lose correct camshaft timing. For instance, if the cylinder head and/or block is planed, the camshaft sprocket/toothed pulley centreline moves closer to the centreline of the crankshaft, which will retard the camshaft timing. If a re-ground camshaft is fitted to an engine and the standard phasing points are lost, the camshaft could be advanced or retarded from the recommended setting.

The way around all of this is to know what the camshaft's correct timing figures are, and to check the camshaft timing and re-adjust if timing proves to be out. The engine should be performance tested at the camshaft manufacturer's recommended setting and then adjusted to 4 degrees retard and then 4 degrees advance. That is three distinct settings and three sets of performance tests. Finally, set the camshaft timing in the position at which the engine runs best. Consider 4 degrees advance or retard to be the limit for any engine.

There are three basic ways that single overhead camshaft engines can have adjustable camshaft timing. The first is by offset keys or dowels if the camshaft and drive sprocket/pulley are located this way. The second is to elongate the hole in the standard sprocket/pulley if the sprocket/pulley is located to the camshaft by dowel. This means that the sprocket/pulley is moveable in relation to the camshaft within limits (length of the slot), and relies on the bolt tension to hold the sprocket in place. If the timing slips it can't slip far and, if the sprocket or pulley comes loose, it will usually be heard as a 'clattering' noise. This is a quick, easy and effective method of converting the standard dowel location of either a sprocket or a pulley to being adjustable within an 8 to 10 degree range. The third method is to buy an aftermarket adjustable vernier sprocket/toothed pulley. These make for very easy camshaft timing adjustments and are recommended if available for your engine.

In all cases, to check or adjust camshaft timing, the valves must be exposed, which will mean removal of the camshaft cover or rocker cover. A dial gauge is positioned on to the valve retainer of the number one cylinder's inlet valve. Note that the inlet full lift

Aftermarket vernier timing wheel fitted to an overhead camshaft engine's cylinder head.

Valve at full lift and the dial gauge needle set to zero.

timing point must be accurately marked on the crankshaft damper/pulley. The crankshaft is turned in its direction of rotation and, when the inlet valve is at full lift (highest reading), the dial gauge needle is set to zero.

Continue to turn the crankshaft in the normal direction of rotation for almost two revolutions until the dial gauge needle reads 0.005in/0.125mm (0.010in/0.25mm can be used instead) from the zero position. Mark the rim of the crankshaft damper/pulley adjacent to the pointer using a fine tipped white marking pen.

Continue to rotate the crankshaft until the 0.005in/0.125mm point (or 0.010in/0.25mm) is reached again on the other side of the full lift point. Stop turning the crankshaft and then mark the damper/pulley adjacent to the pointer. That point midway between the two newly marked points is the 'true' full lift position.

Of course, it may prove necessary to advance or retard the camshaft timing in order to get the precise full lift position to coincide with the camshaft manufacturer's recommended camshaft timing.

Note that on some overhead camshaft engines (not all), the full lift point is very pronounced as a clear peak and, as a consequence, is very easy to find, meaning the camshaft can be timed directly by just going straight to the full lift point and setting the timing at that point. The method of stopping the crankshaft each side of the full lift point and finding and using

CAMSHAFT TIMING PROCEDURE - S.O.H.C. ENGINES

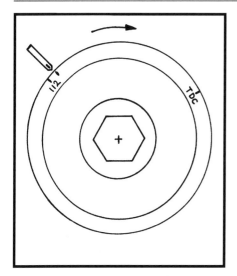

Mark the damper/pulley adjacent to the pointer when the crankshaft is stopped at the chosen position before full lift. The position of the dial gauge needle is what determines the stopping point and, as a consequence, relates camshaft timing to the crankshaft.

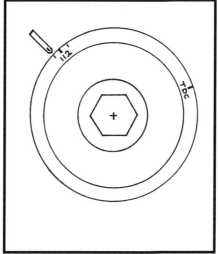

Mark the damper/pulley adjacent the pointer when the crankshaft is stopped at the chosen after full lift point. The dial gauge needle must register the same point as the before full lift timing point.

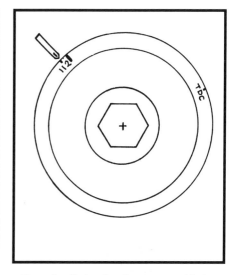

Once the timing has been proved to be correct, the two temporary pulley/damper marks can be removed. If the optimum timing point was found to be at a point different from the cam manufacturer's recommendation, the new timing point can be permanently marked on to the damper/pulley rim.

the mid-point is, however, generally more accurate.

Irrespective of what method is used to time the camshaft, the optimum timing of the camshaft of **any** engine can be found by altering the timing of the camshaft by up to 4 degrees of advance and retard until the 'best' position is found by testing the performance at each setting (including the camshaft manufacturer's recommended setting).

Once the best possible timing point has been found, it can be permanently marked on the crankshaft damper/pulley, or recorded for future use as a certain number of degrees before or after the camshaft manufacturer's specified setting. Alternatively, instead of marking the crankshaft damper or pulley permanently in the first place, the damper/pulley can have the full lift position temporarily marked on it (damper rim painted white and temporary lines scribed on the paint) and only be permanently marked once all the testing and finalising has been done. This method avoids having too many marks on the damper/pulley rim which can get confusing: it is, after all, only the full lift timing point that will be 'moved.'

Visit Veloce on the Web - www.veloce.co.uk

Chapter 8
Camshaft Timing Procedure - T.O.H.C. Engines

TWO VALVE PER CYLINDER ENGINES

The four valve per cylinder twin overhead camshaft engine is universally regarded as being more efficient than its two valve per cylinder counterpart, this is because of the increased valve area at low lift the design offers. Any two valve per cylinder engine, however, responds much the same as any four valve per cylinder engine when it comes to camshaft timing. From this point on, although four valve per cylinder engines are used as examples for camshaft timing, the method of timing the camshafts is identical for both engine types. In fact, when timing a four valve per cylinder engine, only one of the two valves is used for timing purposes which makes it, overall, an identical procedure.

The principle of opening the exhaust valve as late as possible conducive to clearing the cylinder and having the cylinder pressure acting on the piston crown for as long as possible (to produce as much torque as possible) is well-founded. Closing the inlet valve as early as possible conducive to excellent cylinder filling but, obviously, having the inlet valve open long enough, is also a well-founded principle on a two valve per cylinder, twin overhead camshaft engine. This ideal applies equally well to either engine design, it's just that the four valve engine is more efficient and the results are almost always slightly better.

FOUR VALVE PER CYLINDER ENGINES

Virtually all high specification modern production engines feature belt-driven camshafts and four valves per cylinder. The performance capability of these engines is unmatched by other designs on the basis of volumetric efficiency and excellent combustion. What was once the domain of pure racing engines is now, fortunately, available in the majority of modern road cars.

When these engines are modified and tuned correctly and fitted into cars that handle well they are literally brilliant to drive. The engine response

Easily adjustable toothed-belt sprocket. Just undo the four Allen-headed set screws and movement within a fixed range is possible. If the range is not large enough, move the belt a tooth and start again.

is excellent because of good volumetric efficiency (cylinder filling) and they produce amazing power and torque (yes, good torque, especially on the more modern long stroke, long connecting rod, small bore designs) and fuel economy because they are just so efficient. This design of engine gives excellent volumetric efficiency

CAMSHAFT TIMING PROCEDURE - T.O.H.C. ENGINES

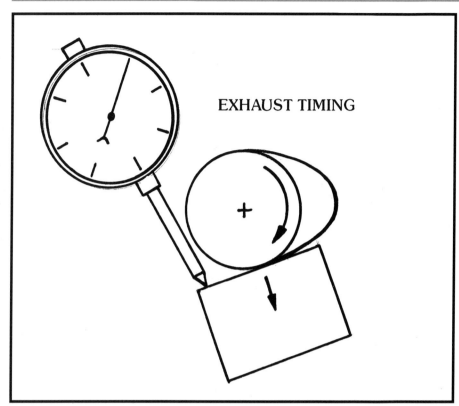

A dial gauge is used to accurately pick the point (in relation to the degree marks on the crankshaft damper/pulley) when the follower starts to move in the case of the exhaust follower, and the point when the follower stops moving in the case of the inlet follower.

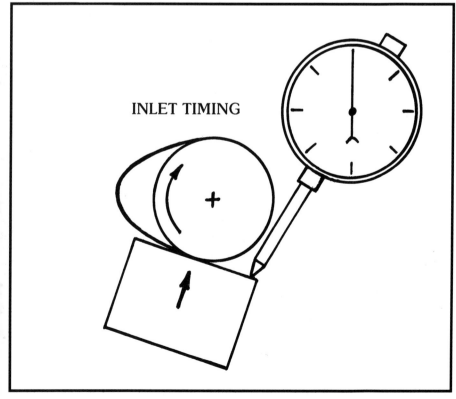

SPEEDPRO SERIES

with very little valve overlap, something many other engine designs cannot do.

With this type of engine the exhaust lobes are all on one camshaft and the inlet lobes are all on another, and the timing of each camshaft is done independently of the other. This allows one camshaft to be moved without moving the other at all, and there is no doubt that the individual camshafts can be positioned for optimum engine performance because of this feature. Most twin camshaft (per bank of cylinders) engines have camshafts that are chain-driven or belt-driven, with some engines being gear driven (this is less common). Offset keys can be used to adjust fixed pulley/gear systems, or the drive pulleys can be replaced with adjustable ones. Chain-driven, twin overhead camshaft engines usually have some form of adjustment as standard equipment, but not always.

There have been some odd designs over the years. Some engines have featured what is, to all intents and purposes, a single overhead camshaft drive but, under the camshaft cover, are two camshafts side-by-side with one of them gear-driven from the 'main' camshaft. Irrespective of this, the gear positions relative to the camshafts are alterable using offset keys. Admittedly, the key/s may have to be adapted from existing available keys. Offset keys are available from camshaft manufacturers and it may be a question of getting a key that is close in size and altering it to suit your application. This is not as difficult as it may sound.

One thing is for sure, when working with engines that can have their camshafts adjusted independently of each other, it is quite easy to see what each camshaft does when one camshaft only is advanced

This dial gauge is set to measure initial follower movement.

or retarded from the standard recommended setting. On these engines, the full lift timing point can be used to set camshaft timing. However, what with variable camshaft lobe accuracy and direct acting of the camshaft onto inverted bucket followers or rockers, it's often much

CAMSHAFT TIMING PROCEDURE - T.O.H.C. ENGINES

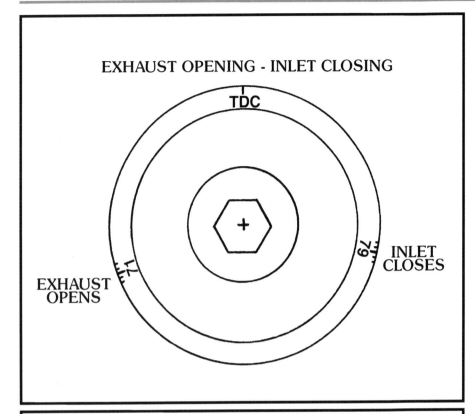

Crankshaft pulley/damper marked with only the camshaft manufacturer's specified exhaust opening and inlet closing points (ignition timing marks are not shown, but they, too, would normally be there). This is the recommended method.

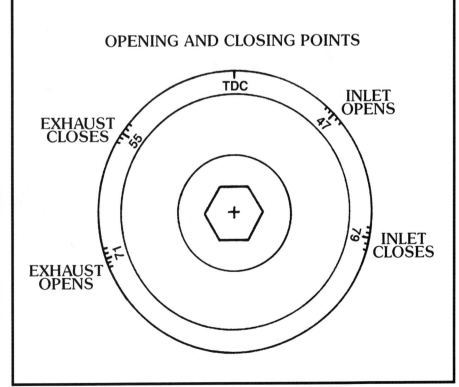

Crankshaft damper/pulley marked with the opening and closing points for inlet and exhaust (ignition timing marks are not shown, but they, too, would normally be there).

SPEEDPRO SERIES

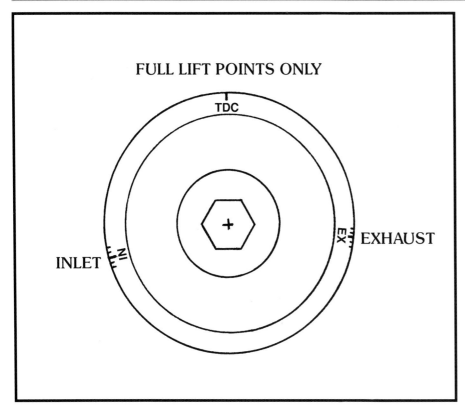

Alternatively, the crankshaft pulley/damper can be marked with full lift positions only (ignition timing marks are not shown, but they, too, would normally be there).

better to make sure that valve clearances are correct and set the camshaft timing (using the first 0.001in/0.025mm of actual valve movement) for exhaust opening and (using the last 0.001in/0.025mm of valve movement) for inlet closing. Forget about the inlet opening point, and the exhaust closing point, and advance or retard each camshaft using the specified inlet closing point and the exhaust valve opening point as datums. The important factor is that the exact opening or closing point is only being used as a datum from which to move the camshaft timing in an effort to find the best timing for optimum engine performance.

While many camshaft manufacturers/grinders recommend using the full lift timing point, power can be lost on these engines if the exhaust valve is opened later than specified (or too early, for that matter). When engines are timed to the manufacturer's specifications, they will always go well, but may not go as well they could: the only way to find out whether it will go better is to advance or retard the timing from the original design setting. It's a good idea to set the camshaft timing at the recommended full lift position and then to check the opening and closing points of the camshafts. If everything is to specification, then it doesn't really matter by which method the camshafts are timed, but, if the opening and closing points are short of the specified opening and closing points, then it does matter. In this situation it is advisable to set **and test** the engine with the camshafts set to the recommended full lift positions, and then test the engine again with the camshafts set at the exhaust opening point and the inlet closing point. Use the settings which produce the best overall engine response for your requirements.

A good reason for using exhaust opening and inlet closing points as datums for timing the camshafts is to nullify the effect of poorly ground camshafts as far as possible. In an ideal world, where every camshaft would be ground perfectly, full lift timing would be ideal as a setting, and then it would be a simple matter of advancing or retarding camshaft timing until the best engine performance was found through testing. However, all too frequently, camshafts are not accurately ground and, by using the full lift timing point, the actual opening point (that first 0.001in.0.025mm of valve movement) can be 3-5 degrees short of the specified opening and closing points of either the exhaust or inlet duration.

A further reason for using the first 0.001in/0.025mm of follower lift for exhaust opening, and the last 0.001in/0.025mm of valve movement is the sometimes experienced difficulty of getting a dial indicator stylus far enough down the follower bore. Some of the smaller four valve per cylinder engines have very small diameter camshaft followers and large camshafts and cylinder head castings.

It is recommended that the crankshaft damper/pulley be marked at the camshaft manufacturer's specified exhaust opening point and inlet closing point of the valves of number one cylinder. The crankshaft damper/pulley is then marked at 2 and 4 degrees after that mark for the exhaust and 2 and 4 degrees before for the inlet. An example is an engine fitted with a camshaft with opening and closing points of 47-79-71-55.

CAMSHAFT TIMING PROCEDURE - T.O.H.C. ENGINES

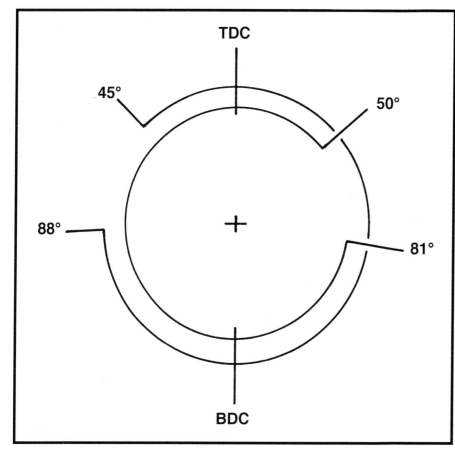

Example timing events.

The exhaust opening point as specified is 71 and the extra degree markings will mean that 67, 69, 73 and 75 degree points BDDC will be marked on the crankshaft damper/pulley. The specified inlet closing point is 79 degrees ABDC, and 75, 77, 81 and 83 degree points ABDC for the inlet closing will need to be marked on the crankshaft damper/pulley.

There are choices with regards to camshaft timing methods but, ultimately, they all lead to the same thing and that is the best possible camshaft timing for your application through testing to see what camshaft timing allows your engine to go best. With this sort of engine - if equipped with a belt and vernier timing pulleys - it is usually very easy to change (minutes) camshaft timing. Re-timing chain or gear-driven camshafts takes more time as everything is enclosed, and the method of altering the camshaft timing is different. If there is no improvement in engine performance at other settings, stay with the camshaft manufacturer's recommended setting, whether it's full lift or inlet closing and exhaust opening points.

The following illustrated examples show a clockwise rotating engine (by far the most common). On an anti-clockwise rotating engine the timing points would be reversed (what is shown on the left would then be on the right). The two lines either side of the main marking are at 2 and 4 degrees.

A full race engine with 9000rpm capability would seldom need to have the exhaust valves opening earlier than 80 degrees BBDC, and a fast road-going engine would seldom have the exhaust valves opening later than 65 degrees BBDC. So, this gives a 15 degree range between the 'wildest' and the 'mildest.' The inlets would not normally be closed later than 80 degrees ABDC and do not really need to be closed earlier than 65 degrees ABDC, which gives a 15 degree range of inlet closing points.

For naturally-aspirated engines, the 80 degree figures are the overall maximums ever to consider before torque and power is lost. This means not worrying too much about the inlet opening points or the exhaust closing points, other than to check for sufficient piston to valve clearance during this phase of the camshafts' cycle.

The camshaft overlap, which is when the exhaust valve is closing and the inlet valve is opening, is a completely separate component of the camshaft timing events. That said, it is, however, advisable to check the durations of camshafts just to see if they are accurate to the manufacturer's listed specifications.

To check/adjust camshaft timing the valve clearances must, of course, be set absolutely accurately (to a tolerance of 0.001in/0.025mm between all lobes).

A dial gauge stylus is placed directly onto the follower of the inlet or exhaust valve of the number one cylinder, and the first 0.001in/0.025mm of follower movement for the exhaust valve opening point, or the last 0.001in/0.025mm of inlet closing movement, is used to set the timing for inlet closing. This timing method is actually very accurate, and will ensure results more in line with the

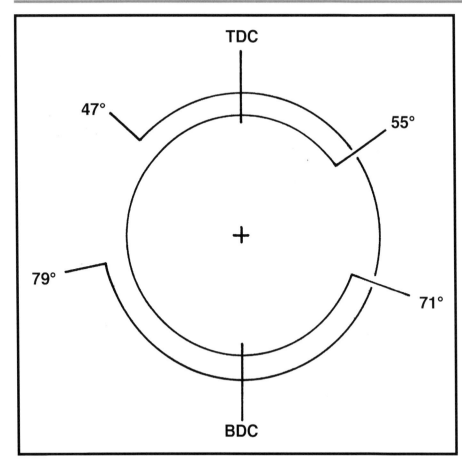

Example timing events.

advertised duration figures normally supplied by the camshaft manufacturer/grinder. The actual overlap degrees may well be down on the advertised figures by a few degrees (3 or 4), but this is nothing to worry about in the overall scheme of things. By all means check the full lift position, or even use it, if preferred. Either way, any measuring point is being used as a datum for further adjustments.

The advantage of twin camshaft valve arrangements, as opposed to single camshaft valve arrangements, is that the camshafts can be moved independently of each other. This is not possible on a single camshaft engine except by the manufacturer/camshaft regrinder at the grinding stage. At the grinding stage this process is called the 'lobe phasing' (relationship between the inlet lobes and the exhaust lobes) and, once ground on to the camshaft, is not alterable (except by regrinding the camshaft).

In many instances the exhaust valve can be opened at 70 to 75 degrees BBDC even on a racing engine turning 9000rpm. This is quite late exhaust valve opening but ideal for producing maximum torque.

Caution! - Always check to see how much piston to valve clearance there is at exhaust closing, as the more the exhaust camshaft is retarded, the more piston to valve clearance is reduced.

What happens with a well-ported four valve per cylinder engine, compared to a well-modified two valve per cylinder engine, is that, with the less efficient two valve per cylinder engine, the exhaust valve just has to be opened earlier than that of a four valve per cylinder engine, in order to clear the cylinder before the piston starts to rise significantly. This means that the cylinder pressure is no longer acting on the piston crown (pushing it down). If the exhaust valve is opened later, the cylinder will not be cleared sufficiently, even after bottom dead centre, and this will hinder the rising piston and reduce power/torque.

A good, real world example of the volumetric efficiency of the four valve per cylinder-type head is a well-modified single overhead camshaft Pinto engine with a modified port standard cylinder head, large valves (44.5mm inlets and 38mm exhausts) and a GP1 camshaft with twin 45mm Weber sidedraught carburettors fitted. A good engine by all accounts, and producing a power curve that started to taper off at about 7500rpm. The camshaft timing was 45-88-81-50. With altered timing (retarded 3 degrees) the exhaust valves opened at approximately 81 degrees BBDC (which is about right for best overall engine performance). The camshaft grinder's recommended timing was 48-85-84-47, but this engine did not go as well when set like this. The same engine was then fitted with a well-ported Sierra Cosworth (four valve per cylinder) cylinder head which had less compression (11:1 instead of 12.5:1), the same carburettors and L1 camshafts timed at 47-79-71-55: the engine then produced excellent power right to 8500rpm (it was never taken beyond this as it was rpm-limited, but the power was not diminishing at 8500rpm - it was improving ...). The

CAMSHAFT TIMING PROCEDURE - T.O.H.C. ENGINES

difference in performance in the same car was dramatic. The Cosworth headed engine outstripped the Pinto-headed engine everywhere, to the point that there really was no comparison. The torque was better because the four valve-headed version of the engine had 10 more degrees of cylinder pressure acting on the piston crown, yet still cleared the cylinder far better than the Pinto-headed version through valve and port efficiency. It is also correct to add that, because of the improved volumetric efficiency of the engine, there was also more air/fuel mixture in the cylinder to burn and produce more power in the first place. This serves to prove the efficiency as, not only was there more gas to expel but it was expelled much more efficiently, even though released 10 degrees later. If the gases were not able to be expelled efficiently, the engine would not have been able to produce the kind of torque/power that it did. Exhaust opening at, approximately, 71 degrees BBDC was right for the rpm range used. When set at 66 degrees the power reduced above 7800rpm but, up until that point, the power was roughly the same, perhaps slightly better, very low down (2000 to 4000rpm). When the engine was timed for 81 degrees of exhaust opening, power and overall engine performance was not as good (less torque).

Visit Veloce on the Web - www.veloce.co.uk

Chapter 9
Engine Testing

Engines can be tested in many ways in the quest for improved performance, but if the results of each adjustment cannot be evaluated, it is most unlikely that much progress will be made. What is required is a testing procedure that is consistent in all ways and readily repeatable. This may seem a difficult task, and it is in certain circumstances. Essentially, what is required is a test that puts the engine under the same load factor each and every time the engine is tested. The camshaft timing can be altered on an experimental basis to improve the characteristics of the engine, and this usually means obtaining a better spread of torque or more torque.

ROLLING ROAD

The rolling road engine dyno is one way of testing engine performance, but there is nothing quite like testing the engine in the environment in which it will be used. If an engine is run up on the rolling road, the information still has to be deciphered and decisions taken as to what (if anything) needs to be changed. To put it bluntly, in the final analysis, you are still left with the situation of the operator having to be a skilled engine tuner. The rolling road is nothing more than a rolling road, it does nothing to tell you what to change in the quest for optimum power/torque from the engine. Just because a car has had its engine tuned on a rolling road, does not mean that maximum power and overall efficiency have been obtained.

Ideas for how to extract maximum power with reliability must come from the engine tuner, not the rolling road. Sure, the rolling road system is nearly always fully equipped - on the basis of having a scientific analysis for CO (air fuel ratio) and an oscilloscope for spark intensity (KVs), and so on. However, it is all very well taking a power reading but someone still has to decide what, if anything, can be improved, what to do about it and then actually do it. This is why many engine tuners do not need a rolling road to get very good results, and also why many rolling road operators do not get as good a result as they like to think they do. Engine tuning is not easy, there are many factors that come into it. Miss one factor and the engine will not go as well as it could. The combination of a very good engine tuner and a correctly calibrated rolling road is hard to beat. The trick appears to be finding the two in one place.

The rolling road in the right hands is an excellent piece of equipment (no doubt about it), but there are variables with rolling roads such as the load simulation for wind resistance. Horsepower readings can be taken but, what is often not understood, can be altered by the flick of a switch (load lightened). Two identical rolling road machines (calibrated correctly) can be made to give two markedly different power readings for the same engine. The point being that, if a rolling road is used for tuning purposes, the same rolling road machine should be used all the time and the settings of the rolling road should be the same each time the car is tested. Taking a car to various rolling roads to see what readings can be got (invariably the

ENGINE TESTING

higher the better) is a complete waste of time. Rolling road operators virtually always give honest appraisals, but what they say is often not what the owners of cars want to hear! Few engines are actually as good as their owners claim, and this has led to problems of overstating engine power. The real losers are the owners of the cars but, apparently, ignorance is bliss.

All too often an engine is tuned on the basis of what the maximum power reading is, but virtually nothing is done to check what torque is being produced in the mid-range. Obtaining maximum power is all very well but not, necessarily, the critical factor if the engine rpm range being used through the gears must span 4000rpm, or so.

TRACK/ROAD TESTING

Another way of testing an engine/car combination is directly on a race track or, to a certain extent, public roads (just depends on what the speed limit is). A lot of useful information can be gathered by testing a road car on the road without excessive speed being involved. Pure racing cars obviously can only successfully be tried and tested on the race track.

Warning! - In no way is speeding on public highways advocated to conduct any tests as described here. There are plenty of sanctioned race tracks which can be hired, disused airfields and many other types of off road sites available for testing purposes without risking life and limb (your own as well as other people's) on public highways.

Warning! - Always carry a fire extinguisher in the car and never test alone. Always have one or two people nearby who can assist immediately should things go wrong. A mobile 'phone is an ideal piece of equipment to have on a test day. It can sometimes be useful to have a helper in the car (non-single-seater cars) with you during testing to use the stopwatch and note the rpm: this allows the driver to concentrate on driving and car control. Ensure that the seats are firmly fixed into the car, that seatbelts are of regulation quality and both driver and passenger are firmly strapped in. Even though this is straightline testing, take no risks.

There can be no better test of engine performance than to check top gear acceleration (starting with the engine just in the power band and then having the engine go from there to just before the power band finishes) over a set distance, and noting the time it took to cover the distance and the rpm that the engine attained at the end of the set distance.

For instance, if an engine's effective power band is known to be from 3500rpm to 7500rpm, the ideal situation would be to approach a set marker cone (road marker cone, for example) on the side of the track at a steady 3500rpm in top gear and, as the car draws level with the first marker, floor the throttle. Just before the power band finishes (say, 7000rpm in this example) the car passes the second set marker cone. At the time the car passes the first marker, the rpm is known (3500rpm) and the stopwatch is started. At the moment the car passes the second marker the stopwatch is stopped and the rpm noted. You may need a few trial runs to set the ideal distance for the second marker. So long as the distance between the markers and the test results are recorded, true comparisons can be drawn in the future by making modifications and duplicating the test. Every time something is changed, the car can be tested: the car will either cover the distance quicker, slower or in the same time. The rpm will either be more, less or the same. Whatever, you will know whether a tuning adjustment has worked or not.

If, after testing over the full distance, you're not sure whether the car is performing better lower down in the rpm range, check the car's acceleration from between 3500rpm to 4500rpm. Simply use a third marker cone set at a shorter distance and use exactly the same testing method. It's fair to say that acceleration can be 'felt' quite accurately by experienced drivers (seat of the pants!) but it is recommended that timed distance testing be used exclusively to gain accurate recordable results.

The set marks can be positioned where it best suits. The fact that the car is in top gear means no gear changes to interfere with the test procedure. Testing, if carried out like this, is as good as it gets because it's simple and because, effectively, the car itself is being utilised as the test bed. There is no guesswork with load factors, or anything else. It is simply a measurement of how effective the engine is at propelling the car.

The golden rule of engine tuning is to do only one thing at a time and test the engine to evaluate that one adjustment. If more than one thing is adjusted at a time, how will you know what did what? It is fair to say that experience comes into this a bit as, if the person tuning the engine has tuned plenty of the same type of engine before, then it is more a case of setting the particular engine to known settings. Even so, a testing procedure like this is still valid as it serves to give some sort of comparison, such as 'the other engine went a bit better than this in a similar weight car' or 'this engine goes better than the other engine, but this car is 200 pounds lighter and lower geared,' and so on.

DRAG STRIP

The drag strip is another venue that certainly shows up any deficiencies. Any car that is going to be competitive in drag racing just has to have an engine that is good everywhere in its rev range. The point being that there is a set distance and the race is against the clock. If you do lose time for any reason (missed shift, for instance), you cannot gain it back. Overall, the driver has to be very skilled at getting away and gear changing. This is not as good a test as the top gear only test, because there are variables, such as the gearchanging, meaning it may not be a consistent test. Any faults are highlighted as slow times in comparison to other similar cars.

Adjustments to camshaft timing can be tried (maximum of 4 degrees retarded or advanced) and tested. It does not matter what the final setting is, as long as you know that you have set the engine to where it runs best for your application.

On pushrod overhead valve engines in particular, making camshaft timing adjustments and testing them can be a time-consuming job as most timing chain covers are not all that easy to get off, and can be difficult to re-seal against oil leaks. Modification of the timing chain cover is often possible so that a plate can be removed and the camshaft sprocket/gear easily got at and adjusted. If the cover is well designed, there is no reason why a camshaft cannot be adjusted and the car ready to go again in 20 to 30 minutes, which is useful in a drag racing situation.

The next quarter mile time and speed will tell you when you have made the right adjustments, but driver consistency is equally important.

ENGINE DYNO

Another way of testing an engine is on an engine dyno. These devices offer good engine-to-engine comparison information, and also comparison for fault-finding. There is no doubt that an engine tested and tuned correctly on an engine dyno will be spot-on in the hands of a good engine tuner/operator.

Unfortunately, for a variety of reasons, many engines which have been tuned on engine dynos fail to give the expected performance in service. It can be that the engine was not correctly loaded during testing, or the water cooling and fuel supply arrangements were markedly different to those available when the engine was installed in the car, making the results worthless.

Fitting an engine onto an engine dyno is not a five minute job. The engine has to be taken out of the car and, possibly, transported to the engine dyno facility, fitted into the machine, run up and tested, adjusted, removed from the machine, re-installed in the car and tested again in the car. There is considerable expense involved and sometimes the rewards are meagre. The usual situation is that, when an engine is built, it is tested and tuned by the engine builder so that everything about the engine is a known quantity before it leaves the engine builder's premises. Power production is proved and the reliability of the engine is proved (to a degree).

ALSO FROM VELOCE PUBLISHING -

HOW TO BUILD, MODIFY & POWER TUNE CYLINDER HEADS
- 2nd Edition
by PETER BURGESS & DAVID GOLLAN

ISBN 1 901295 45 1
Price £14.99 *

A book in the *SpeedPro* series.

- The complete practical guide to successfully modifying cylinder heads for maximum power, economy and reliability.
- Understandable language and clear illustrations.
- Avoids wasting money on modifications that don't work.
- Applies to almost every car/motorcycle (does not apply to 2-stroke engines).
- Applies to road and track applications.
- Peter Burgess is a professional race engine builder. David Gollan B ENG HONS is a professional engineer.

CONTENTS
Horsepower and torque • Equipment and tools • Airflow • Building a flowbench • Porting • Valves, valve seats and inserts, valve guides, valve springs • Unleaded fuel conversion • Fuel economy • Includes an overview of camshafts, exhaust systems, ignition, cylinder blocks • Case studies of a selection of popular cylinder heads • Index.

SPECIFICATION
Paperback. 250 x 207mm (portrait). 112 pages. Over 150 black and white photographs/line illustrations.

RETAIL SALES
Veloce books are stocked by or can be ordered from bookshops and specialist mail order companies. Alternatively, Veloce can supply direct (credit cards accepted).

* *Price subject to change.*

Veloce Publishing Plc, 33 Trinity Street, Dorchester, Dorset DT1 1TT, England. Tel: 01305 260068/ Fax: 01305 268864.

SPEEDPRO SERIES

ALSO FROM VELOCE PUBLISHING -

HOW TO BUILD & POWER TUNE WEBER & DELLORTO DCOE & DHLA CARBURETORS
- 2ND Edition
by Des Hammill

ISBN 1 901295 64 8
Price £14.99*

A book in the **SpeedPro** series. All you could want to know about the world's most famous and popular high-performance sidedraught carburetors. Strip & rebuild. Tuning. Jetting. Choke sizes. Application formula gives the right set-up for *your* car. Covers all Weber DCOE & Dellorto DHLA carburetors.

CONTENTS
COMPONENT IDENTIFICATION: The anatomy of DCOE & DHLA carburetors • DISMANTLING: Step-by-step advice on dismantling. Assessing component serviceability • DIFFICULT PROCEDURES: Expert advice on overcoming common problems in mechanical procedure • ASSEMBLY: Step-by-step advice on assembly. Fuel filters. Ram tubes. Fuel pressure • SETTING UP: Choosing the right jets and chokes to get the best performance from *your* engine • FITTING CARBURETORS & SYNCHRONISATION: Covers alignment with manifold and balancing airflow • FINAL TESTING & ADJUSTMENTS: Dyno and road testing. Solving low rpm problems. Solving high rpm problems. Re-tuning.

THE AUTHOR
Des Hammill has a background in precision engineering and considers his ability to work very accurately a prime asset. Des has vast experience of building racing engines on a professional basis and really does know how to get the most out of a Weber or Dellorto carburetor. Having lived and worked in many countries around the world, Des currently splits his time between the UK and New Zealand.

SPECIFICATION
Softback • 250 x 207mm (portrait) • 112 pages • Over 140 black & white photographs and line illustrations.

RETAIL SALES
Veloce books are stocked by or can be ordered from bookshops and specialist mail order companies. Alternatively, Veloce can supply direct (credit cards accepted).

* *Price subject to change.*

Veloce Publishing Plc, 33 Trinity Street, Dorchester, Dorset DT1 1TT, England. Tel: 01305 260068/ Fax: 01305 268864.

Visit Veloce on the Web - www.veloce.co.uk

ALSO FROM VELOCE PUBLISHING -

HOW TO BUILD & POWER TUNE DISTRIBUTOR-TYPE IGNITION SYSTEMS
by Des Hammill

ISBN 1 874105 76 6
Price £9.99*

A book in the **SpeedPro** series. Expert practical advice from an experienced race engine builder on how to build an ignition system that delivers maximum power reliably. A lot of rubbish is talked about ignition systems and there's a bewildering choice of expensive aftermarket parts which all claim to deliver more power. Des Hammill cuts through the myth and hyperbole and tells readers what *really* works, so that they can build an excellent system without wasting money on parts and systems that simply don't deliver.

Ignition timing and advance curves for modified engines is another minefield for the inexperienced, but Des uses his expert knowledge to tell readers how to optimise the ignition timing of *any* high-performance engine.

The book applies to all four-stroke gasoline/petrol engines with distributor-type ignition systems, including those using electronic ignition modules: it does not cover engines controlled by ECUs (electronic control units).

CONTENTS
Why modified engines need more idle speed advance • Static idle speed advance setting • Estimating total advance settings • Vacuum advance • Ignition timing marks • Distributor basics • Altering rate of advance • Setting total advance • Quality of spark •

THE AUTHOR
Des Hammill has a background in precision engineering and considers his ability to work very accurately a prime asset. Des has vast experience of building racing engines on a professional basis. Having lived and worked in many countries around the world, he currently splits his time between the UK and New Zealand.

SPECIFICATION
Softback • 250 x 207mm (portrait) • 64 pages • Over 70 black & white photographs and line illustrations.

RETAIL SALES
Veloce books are stocked by or can be ordered from bookshops and specialist mail order companies. Alternatively, Veloce can supply direct (credit cards accepted).

* *Price subject to change.*

Veloce Publishing Plc, 33 Trinity Street, Dorchester, Dorset DT1 1TT, England. Tel: 01305 260068/ Fax: 01305 268864.

SPEEDPRO SERIES

ALSO FROM VELOCE PUBLISHING -

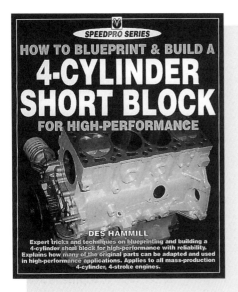

A book in the **SpeedPro** series.

- Applies to all 4-cylinder car engines (except diesel & two-stroke).
- Essential reading for millions of car owners looking for more power.
- Expert advice in non-technical English accompanied by clear photos & line illustrations.
- Saves money by eliminating techniques that don't work and by maximising the use of standard components.
- Written by a professional competition engine builder.

HOW TO BLUEPRINT & BUILD A 4-CYLINDER SHORT BLOCK FOR HIGH PERFORMANCE
by DES HAMMILL

ISBN 1 874105 85 5
Price £13.99 *

CONTENTS
A complete practical guide on how to blueprint (optimize all aspects of specification) any 4-cylinder, four-stroke engine's short block to obtain maximum performance and reliability without wasting money on over-specced parts. Includes choosing components, crankshaft & conrod bearings, cylinder block, connecting rods, pistons, piston to valve clearances, camshaft, engine balancing, timing gear, lubrication system, professional check-build procedures and much more. Index.

SPECIFICATION
Paperback. 250 X 207mm (portrait). 112 pages. Around 200 black & white photographs/illustrations.

RETAIL SALES
Veloce books are stocked by or can be ordered from bookshops and specialist mail order companies. Alternatively, Veloce can supply direct (credit cards accepted).

* *Price subject to change.*

Veloce Publishing Plc, 33 Trinity Street, Dorchester, Dorset DT1 1TT, England. Tel: 01305 260068/ Fax: 01305 268864.

Visit Veloce on the Web - www.veloce.co.uk